D1105929

Advances in
PARASITOLOGY

VOLUME 19

Advances in
PARASITOLOGY

Edited by

W. H. R. LUMSDEN (Senior Editor)

*University of Dundee, Department of Animal Services, Ninewells Hospital and
Medical School, Dundee, Scotland*

R. MULLER

Commonwealth Institute of Helminthology St. Albans, England

and

J. R. BAKER

*NERC Culture Centre of Algae and Protozoa, Institute of Terrestrial Ecology, Cambridge,
England*

VOLUME 19

1981

ACADEMIC PRESS

A Subsidiary of Harcourt Brace Jovanovich, Publishers

London   New York   San Francisco   Toronto   Sydney

ACADEMIC PRESS INC. (LONDON) LTD
24/28 Oval Road,
London NW1 7DX

United States Edition published by
ACADEMIC PRESS INC.
111 Fifth Avenue,
New York, New York 10003

Copyright © 1981 by
ACADEMIC PRESS INC. (LONDON) LTD.

All Rights Reserved
No part of this book may be reproduced in any form by photostat, microfilm, or any other
means, without written permission from the publishers

British Library Cataloguing in Publication Data

Advances in parasitology.
 Vol. 19
 1. Veterinary parasitology
 591.2'3   SF810.A3

  ISBN 0-12-031719-2
  LCCCN 62-22124

Typeset by Bath Typesetting Ltd., Bath
and printed in Great Britain by
John Wright & Sons (Printing) Ltd., at The Stonebridge Press, Bristol

CONTRIBUTORS TO VOLUME 19

D. W. T. CROMPTON, *Molteno Institute, University of Cambridge, Downing Street, Cambridge CB2 3EE, England*

Z. KABATA, *Department of Fisheries and Oceans, Pacific Biological Station, Nanaimo, B.C., Canada V9R 5K6*

JOHN S. MACKIEWICZ, *Department of Biological Sciences, State University of New York at Albany, Albany, New York 12222, USA*

V. R. PARSHAD, *Department of Zoology, Punjab Agricultural University, Ludhiana 141004, India*

FEB 2 5 1982

PREFACE

For Volume 19 of *Advances in Parasitology* there is little to add prefatorially to what we said for Volume 18; but we should reiterate the statement of policy which we made in that volume. It is: that we wish to view parasitology in its widest sense, taking the parasitic mode of life as the unifying principle, and so admit for review *any* organism group whose members follow that life style; that we should try to select for review areas of parasitology which have in fact advanced significantly at the time of their being reviewed. Considering that our own view of parasitology—as all 'medical' parasitologists—is comparatively restricted, we shall welcome suggestions of any subjects ripe for review on the principles stated.

1981

W. H. R. Lumsden
J. R. Baker
R. Muller

CONTENTS

Copepoda (Crustacea) Parasitic on Fishes: Problems and Perspectives

Z. KABATA

Aspects of Acanthocephalan Reproduction

V. R. PARSHAD AND D. W. T. CROMPTON

Caryophyllidea (Cestoidea): Evolution and Classification

JOHN S. MACKIEWICZ

Copepoda (Crustacea) Parasitic on Fishes: Problems and Perspectives

Z. KABATA

*Department of Fisheries and Oceans, Pacific Biological Station,
Nanaimo, B.C., Canada V9R 5K6*

I. INTRODUCTION

The unceasing progress of science moves in an uncoordinated manner, more often than not following the lines of least resistance, or exploiting opportunities provided by technological breakthroughs. Stimuli are indiscriminately provided to various fields of science by new concepts or theories. Rapid progress on narrow fronts leaves behind lacunae of ignorance. It is advisable, therefore, to pause occasionally and to take stock, to marshall our achievements, assess our failures, and—hopefully—plot rational paths ahead.

This review is intended to serve at least some of these purposes. It cannot lay claim to completeness and, like all reviews, is circumscribed by the limitations and biases of its author. Nonetheless, it might provide sufficient food for thought to those interested in parasitic Copepoda and thus stimulate further progress in the field. It was originally intended to cover Copepoda parasitic on invertebrates, as well as those living on fishes. However, a recent excellent review (Gotto, 1979) dealt with the former more than adequately. I have, therefore, limited my scope to the latter, taking as starting point the most recent review of similar type, published by Bocquet and Stock (1963). Almost two decades have elapsed since that work appeared, and a fresh look at the field is in order. Events and reports predating Bocquet and Stock's review are omitted here, except when it is necessary to mention them in the context of more recent developments.

Parasitic Copepoda of fishes, especially those of marine fishes, are far from being adequately known. Descriptions of new species and genera, even of families, keep appearing in the literature and influencing our views on the order as a whole, on the relationships within it, and on the nature of both the parasites and parasitism itself. An eminent biologist, who once said to me "surely all this was done in the 19th Century," was rather less than right. There is still a great deal of new descriptive work to be done. Much of the old work must also be repeated, with rigorous application of new concepts and standards. The systematist's work is far from finished. In this work he has now a new and powerful tool, the scanning electron microscope. This splendid aid to systematic work was *in statu nascendi* when Bocquet and Stock's review was published in 1963. Now it can be used, though its use is still beset with technical difficulties. I have had my share of frustrations resulting from attempts to use it on parasitic Copepoda. The greatest difficulty is in preparing specimens for examination, particularly in dessicating them without distortion. We are still seeking a reliable and easy method of drying these arthropods. Another serious difficulty is the removal of contaminating fish mucus, which often escapes detection under the light microscope, only to appear as a thick layer obscuring important structures when the specimen is observed under the higher magnification of the scanner.

The general progress in knowledge of the biology of arthropods has now begun to affect the field of parasitic copepods. The last two decades have brought evidence that these parasites are now being looked at with increasing frequency as living organisms, and that their biology, not just their structure, is beginning to fall under the scrutiny of the researcher. Physiology and host–parasite relationships are being studied, though here also many technical problems hamper the investigations. This is particularly true of the parasites of marine fishes, not very amenable to experimental treatment. I am reminded of an eager student who wished to study the development and life cycle of a lernaeopodid copepod living on a man-eating shark, and who had to be reminded of the possible consequences! We are still waiting for a compendium on the biology of parasitic Copepoda. More than anything else, it would help us to see them as functioning animals and to point out directions for future studies.

The second technological breakthrough of great importance to the study of parasitic copepods is the computer, and the flourishing development of mathematical methods in general. With their aid, we can quantify knowledge, and extract new information from data long available but never fully analysed. The ecological parasitologist and the student of population dynamics are now able to cut through appearances and see many aspects of the copepod's biology in their true light. There are difficulties here, too. There is still the need to balance the expertise of the "number cruncher" with sufficient biological insights, a task not among the easiest.

All in all, the next two decades promise substantial progress in our understanding of parasitic Copepoda. I envisage this progress as being mainly in the field of physiology and ecology, the study of life cycles and host–parasite relationships. An improvement in our knowledge of systematics will come as a "spin-off" from these activities, as well as from more directly applied studies.

This review is put together in the hope that it will serve as a convenient focus. The field is divided into topics, beginning with the more traditional subjects of anatomy and morphology, systematics and classification, and less traditional evolution and phylogeny. Then follow reviews of new interesting forms described during the last two decades, a look at the life cycles, host–parasite relationships and some aspects of the biology of parasitic copepods.

II. MORPHOLOGY AND ANATOMY

The study of parasitic Copepoda is a broad field, with many facets. Inasmuch as it has not quite outgrown its descriptive stage, it still tends to be dominated by descriptions of newly discovered species. In spite of their preoccupation with such discoveries, students of parasitic copepods have long been curious about the structure of the animals they were describing. Admirable attempts to study the anatomy of individual species date back more than a century and a half. Some of them, e.g. the illustration of the anatomy of *Actheres percarum* Nordmann, 1832 by its discoverer (Nordmann, 1832), have been good enough to merit repeated reproductions for many decades. Nevertheless, the study of anatomy has never been high in priority, so that we still have only very vague ideas of the internal structure of parasitic Copepoda.

In the course of the last 20 years, little has been done to improve this situation. Scattered and fragmentary papers on anatomy and morphology, some of excellent quality, have been published during that time. Vaissière (1961) described the structure of the ocelli in several species of copepods (*Caligus diaphanus* Nordmann, 1832; *Lepeophtheirus nordmanni* (Edwards, 1840); *Pennella* sp.; *P. varians* Steenstrup and Lütken, 1861; *Pandarus* and *Nessipus*), with very good illustrations based on reconstruction from serial sections. A similar method was used by Einszporn (1965a) to examine the structure of the alimentary canal of *Ergasilus sieboldi* Nordmann, 1832. On the basis of histological differences, she divided this organ into three parts: oesophagus, midgut (mesodeum) and hindgut (proctodeum). There were no sharp histological borders between these parts. Mesodeum was "digestive and absorbent," with its surface expanded by diverticula and equipped with several types of secretory cells. Einszporn was interested in the food and mode of feeding of *E. sieboldi* and devoted much attention to food remains in the lumen of the digestive tract. Kabata and Cousens (1972), in the course of their study of the attachment of Lernaeopodidae, discovered three types of internal structure (independent of the shape) of the bulla, the anchoring organ unique to that family of copepods. The first type occurs in freshwater species of the family and is characterized by the presence of a large canal, looping through the bulla and opening through the manubrium via two parallel, narrow ducts. The second type is typical of all marine lernaeopodids living on teleost fishes. In this type, the ducts run through the manubrium and break up into numerous fine ducts within the anchor of the bulla. The pattern of these ducts was used by Leigh-Sharpe (1925a) to establish several new species of the genus *Clavella*. Kabata (1963a) demonstrated that the patterns are variable and cannot be used as a taxonomic criterion. The third type occurs in Lernaeopodidae parasitic on elasmobranch hosts. The parallel ducts of the manubrium end blindly in the matrix of the anchor, which is either granular or fibrous. The ducts of the manubrium in all three types connect with ducts running the entire length of the second maxillae. Cousens (1977) studied the structure of these ducts. They proved to be voluminous in their proximal parts and ended in blind sacs near the bases of the second maxillae, projecting somewhat into the lumen of the trunk. Each has a small, diffusely structured gland connected with it by a very narrow duct (Cousens's "proximal maxillaryduct"). Together with the bulla, the ducts form a closed system filled with fluid, probably secreted by the glands. The fluids can be moved within the system from one maxilla to the other, and back, by repeated alternate contractions of these appendages. The passage through the bulla provides the opportunity for the fluid (the nature of which is still unknown) to enter the tissues of the host. Although no structural evidence for the existence of such exchanges was uncovered, Cousens demonstrated that the bulla could pick up some substances from the host (see p. 57).

Electron microscopy was employed to study the structure of the reproductive apparatus of parasitic copepods. Manier *et al.* (1977) investigated the male reproductive system of *Naobranchia cygniformis* Hesse, 1863, while

Rousset *et al.* (1978) made a thorough study of the same system in both sexes of *Chondracanthus angustatus* Heller, 1865. Both papers dealt also with spermatogenesis.

Kabata (1974a) studied the functional morphology of the mouth tube of *Caligus* and, on the basis of its structure, suggested the mode of feeding of this copepod (see p. 56).

As can be seen, no systematic effort has been made to study the functional morphology and anatomy of parasitic copepods. Perhaps the broadest review of this subject is the one given by Kabata (1979). It too, however, deals with only one aspect of the problem, that of the impact of parasitism on copepod morphology. More is needed.

Studies of morphology sometimes bring in their train phylogenetic reflections. The origin of an appendage and its morphological affinities are no less interesting than its structure and function. Parasitic copepods become modified, under the influence of adaptation to novel modes of life, to the point at which it is often difficult to judge their morphology in phylogenetic perspective. One of the best known controversies in this field is the so-called "Heegaardian heresy." Heegaard (1947) postulated the existence in parasitic Copepoda (except for Caligidae and some others) of a single pair of maxillae and two pairs of maxillipeds. His views were counter to the more orthodox acceptance of two pairs of maxillae and one of maxillipeds. An additional pair of appendages had to be inserted into the segmental series in Caligidae. Bocquet and Stock (1963) in their review supported the opposite side of the controversy. Lewis (1969) carefully considered the morphology of the post-antennary process (the name he himself proposed), which Heegaard interpreted as the first maxilla, and concluded that no part of this structure is, or has been, associated with any appendage. I have summarized and reviewed the present status of this controversy (Kabata, 1979). It is not yet resolved, although the balance of evidence seems to be against Heegaard's view on the homology of copepod appendages. Two ways are open to those who wish to contribute to the solution of this interesting theoretical problem. The first one is the study of the ontogeny, particularly of its early stages. The main difficulty with this approach is the fact that the change from the last nauplius to the copepodid stage is a veritable jump over an organogenetic abyss. Whereas the naupliar appendages series ends with the mandibles, the copepodid usually emerges from the moult with all its appendages in position. In some species of Caligidae, the postantennary process appears in ontogeny together with the appendage that is accepted as the first maxilla. There is no reason at present to consider the postantennary process as anything but a specialized structure, secondarily evolved as an aid to prehension. It is still possible, though not very probable, that the ontogeny of more primitive copepods retains some features which would force us to reassess our views on the first maxilla and "second maxilliped." The second line of investigation is the study of the nervous system and the way in which it supplies appendages that have allegedly lost their place in the segmental sequence. Lewis's (1961) study of the nervous system of *Lepeophtheirus dissimulatus* Wilson, 1905, is too superficial to allow any conclusions. The same is true of earlier work on this subject (Ferris and Henry, 1949; A. Scott, 1901). We need a very thorough study of the morphology and histology of the central nervous system of a large copepod, such as *Cecrops latreilli* Leach, 1816. The course of all nerves should be carefully followed and their origin studied in histological sections. If innervation can be accepted as a clue to the homology of appendages, a comparison of the nervous system of *C. latreilli*, which has no sternal furca, with one of a species possessing such a furca, might also throw some light on the nature of that enigmatic structure.

Lewis (1966) saw the sternal furca as a remnant of a system of interpodal bars provided with stylets. He believed that in primitive forms there existed a series of plates, or bars, one at the anterior end of each pedigerous segment. Each of these plates had a furca-like structure in the centre. (Interpodal bars with stylets, though not central in position, still exist in some Eudactylinidae and Kroyeriidae.) Most, or all, of these structures have become reduced, vestiges only remaining in, for example, *Euryphorus* or *Nessipus*. The furca of modern Caligidae and their relatives is derived, Lewis suggested, from the medial process of the interpodal bar which once extended between the bases of the maxillipeds. It has been retained and developed because it has assumed a new function. Lewis's plausible explanation, to become incontrovertible, must be supported by morphological and biological data. We have never been able to find an interpodal bar between the maxillipeds. As a remnant of the ancestral condition, perhaps it will be discovered in a primitive copepod, or in an early stage in the ontogeny of such a copepod. The function of the furca is still a matter for conjecture. The fact that some caligids successfully exist without it makes its purpose difficult to understand. Perhaps experiments with the removal of the furca would enable us to determine its survival value to the copepod. Until we have found an interpodal bar between the maxillipeds and until we determine the true function of the furca, we must view Lewis's explanation as hypothetical.

Finally, recent years have seen the clarification of the homology of the structures previously commonly designated as the caudal rami, or caudal furca. Having reviewed the caudal appendages in various groups of Crustacea, Bowman (1971), referring to Copepoda, concluded "Since the caudal rami are the appendages of the anal somite, they are uropods." I support the use of that term as having morphological significance, often obscured in parasitic copepods, particularly those that undergo extensive metamorphosis. Adult females of Pennellidae sometimes lose uropods altogether. In Lernaeopodidae they have turned into digitiform posterior processes, the nature of which has not always been understood. Nunes-Ruivo (1966) suggested that the posterior processes are uropods. Kabata and Gusev (1966) traced the progress of maturation in the female *Eubrachiella gaini* (Quidor, 1913) and observed the gradual transformation of the recognizable, armed uropods into tubercles which no longer bear any resemblance to the original structures. It should be noted that the lernaeopodid uropods, which sometimes attain enormous size (in *Dendrapta cameroni longiclavata* Kabata and Gusev, 1966, they are longer than the rest of the body), are often accompanied by other digitiform structures on the posterior margin of the trunk. In spite of their resemblance to the uropods, these structures are secondary trunk modifications and have no morphological homologues.

III. CLASSIFICATION AND PHYLOGENY

Bocquet and Stock (1963), discussing the phylogeny of parasitic copepods, expressed the view that we often know more about the relationships between major taxa than we do about those between genera in a family, or between species in a genus. This was a rather sweeping statement. While it is true that our knowledge of intrafamilial and intrageneric relationships is scanty, our understanding of affinities between families or suborders is no better. We are still employing a classification set up by Sars (1903) on the "pigeon hole" principle and, for parasitic Copepoda, elaborated by Wilson (1910) on the basis of his "testimony of degeneration."

I attempted to review the classification of Copepoda with the aid of "primary" or "primitive" characteristics (Kabata, 1979) and ascribed to the "secondary" characteristics much less importance than was accorded to them by Wilson. My proposed new classification of Copepoda will be outlined below but it is worthwhile here to recapitulate some of the arguments in favour of the "primary" characteristics as basic clues to the phylogeny of copepods, parasitic copepods in particular. In essence, these characteristics were preferred because they were deemed less susceptible to the influence of parallelism and convergence, often imposed by parasitism on copepod morphology; the latter should constitute the basis for classification. The use of the "testimony of degeneration" (i.e. of the secondary characteristics) has resulted in the establishment of such patently artificial taxa as Lernaeopodoida, and in forcing together siphonostomatous, poecilostomatous and gnathostomatous species in the same suborder, Cyclopoida. There remains the question of how to determine whether a given characteristic is primary or secondary. I know no easy answer to that question. Generally speaking, the primary characteristics form the morphological matrix inherited from the ancestral species and recognizably retained by the copepod. For example, I consider the gnathostome type of mouth-parts to be primitive, because it is widespread throughout Copepoda (see g, Fig. 3) and because its occurrence is not determined by the mode of life. Pelagic Calanoida, demersal Harpacticoida, predatory and saprophytic species, all may be equipped with mouth-parts of similar type, suggesting ancestral affinity. On the other hand, species with similar modes of life and a high degree of morphological resemblance may have mouth-parts of different types. This would suggest convergent evolution of two distinct stocks, rather than phylogenetic affinity. As with many other problems of classification, this one has to be considered on the merits in each individual instance, and judged in a broad context. This leads to some degree of subjectivity, deplored by many, and once more brings forth the comment that systematics is a science which is also an art. Though aware that my views in this instance are open to challenge and might prove inappropriate for some systematic problems, I am confident that they are valid in principle. More detailed discussion is given by Kabata (1979).

Attempts to give a phylogenetic interpretation to copepod systematics are not new. They are, however, rare. In the course of the last two decades some, differing in scope and detail, have been recorded. Markevich (1964) published a brief account of the phylogenetic relationships of Copepoda parasitica, based on a paper delivered at the XVI Congress of Zoology in 1963. In spite of its brevity, it covered a wide field. Markevich derived the majority of "Copepoda parasitica" from the free-living Cyclopoida. He rejected as artificial the suborder Lernaeopodoida; he did not reject Notodelphyoida, although he placed them close to Cyclopoida. The point of contact between Caligoida and Cyclopoida was in the affinity of Caligidae to Taeniacanthidae, through the genus *Assecula* Gurney, 1927, a member of the latter family. This affinity was, unfortunately, not discussed. The validity of *Assecula* was discounted by A. Scott (1929) and the entire argument appears to lack a sufficiently solid foundation. Markevich rejected Wilson's (1932) splitting of the four caligid subfamilies into independent families: Caligidae, Eury-phoridae, Pandaridae and Cecropidae. In so doing, he went counter to the views generally prevailing. An interesting discussion of Sphyriidae brought out the fact that their affinities are obscure. The similarities between sphyriid and lernaeopodid males were weighed against similarities between sphyriid females and some female chondracanthids. The question was left unanswered. I believe that the morphology of the males points unmistakably to the rela-tionship between Sphyriidae and Lernaeopodidae and hence to the place of the former in the suborder Siphonostomatoida.

The scope of Fryer's (1968) paper was more modest, but his treatment of the subject more detailed. He discussed the evolution and adaptive radiation of African Lernaeidae and gave a graphic representation of this process, which has been simplified here in Fig. 1. He believed, justifiably, that *Lam-proglena* Nordmann, 1832, with its partially retained segmentation and tagmosis, musculature and absence of the holdfast characteristic of other Lernaeidae, is a rather primitive genus, derived as a separate branch from the parental lernaeid stock. Somewhat aberrant *Lamproglenoides* Fryer, 1964, and elongated *Afrolernaea* Fryer, 1956, are offshoots of that branch. The former might be split from the line that led to *Lamproglena cornuta* Fryer, 1965; the latter is closest to *Lamproglena intercedens* Fryer, 1965. The apparent tendency of *Lamproglena* to produce ever more elongated species with a concentration of appendages on the anterior end, foreshadows the mor-phological condition prevailing among other lernaeid genera. *Lamproglena*, *Lamproglenoides* and *Afrolernaea* are characterized by the possession of uniserial egg sacs, not usual among cyclopoid copepods. In Fryer's view, *Lernaea* evolved as a separate branch from the ancestral stock. In Africa at least, it produced six adaptive radii, two of which radiated again to produce more than one species. The third branch has much in common with *Lernaea* and might have been ancestrally close to it. It comprises *Opistholernaea*

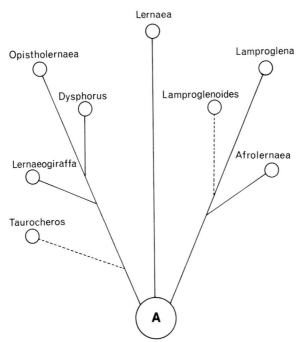

Fɪɢ. 1. Hypothetical phylogenetic tree of African Lernaeidae. (Modified from Fryer, 1968.)

Yin, 1960; *Dysphorus* Kurtz, 1924; and *Lernaeogiraffa* Zimmermann, 1922. For comparative purposes, Fryer included here also the little-known South American *Taurocheros* Brian, 1924. All resemble *Lernaea* in the type of metamorphosis and in host–parasite relationships. *Opistholernaea* appears the most ancient of them and has branched out into several species. Fryer's views, based on the consideration of morphology, biology and zoogeography, are a valuable contribution to our knowledge of the phylogeny of copepods within a single family. His methods should be followed in the conduct of similar exercises.

Ho (1978) published an abstract of a paper to be given at the IV International Congress of Parasitology, dealing with the origin of Chondracanthidae. While deficient in detail by its very nature, the abstract proposed that Chondracanthidae are derived from a common stock with poecilostome Sabelliphilidae, parasitic on invertebrates. In particular, affinity was postulated between Chondracanthidae and four sabelliphilid genera parasitic on holothurians (*Calypsarion* Humes and Ho, 1969; *Calypsina* Humes and Stock, 1972; *Caribulus* Humes and Stock, 1972; and *Scambicornus* Heegaard, 1944). Ho stated "that the common ancestral stock of these sabelliphilid genera had exploited the opportunity of life on the benthic fish hosts [which] occurred in the same community with their holothurian hosts and thus gave rise to Chondracanthidae." Parasitism is seen, then, by Ho as a tool of evolution.

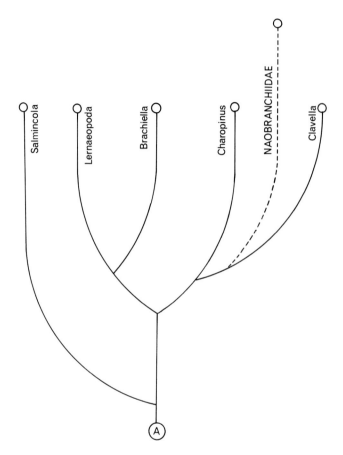

FIG. 2. Hypothetical phylogenetic tree of Lernaeopodidae. (Modified from Kabata, 1979.)

I put together a phylogenetic tree of Lernaeopodidae (Kabata, 1979). This concept, which developed over the years, was first presented during the First International Congress of Parasitology in 1964 and published in abstract (Kabata, 1966). Briefly, I believe that Lernaeopodidae evolved in the unspeci-fied past from the ancestral stock by developing permanent fusion of the second maxillae and the bulla, the successor of the larval frontal filament. During their early history, a branch, referred to as the *Salmincola*-branch, split off from the main stem of the family (Fig. 2). The stem eventually divided into four branches: *Lernaeopoda*, *Charopinus*, *Brachiella* and *Clavella*. G. Fryer (personal communication) doubted whether the suggestion of a three-way split of the main stem (shown in Kabata, 1979, p. 48) would be accepted by those who "deal in dichotomies." Fryer's point is well taken, although family trees quite often show multiple divisions to indicate adaptive radiation at the specific level. A separation of the *Charopinus*-branch from the *Lernaeopoda*-branch (the former characterized by a dorsal position of the posterior processes of the trunk, as opposed to the ventral position of those processes in the latter) is a clear dichotomy. I suggested that the *Clavella*-branch, the third line of the triple split, has some affinities with the *Charopinus*-branch, as shown by the dorsal position of vestigial processes occurring on the trunks of some species of *Clavellisa* Wilson, 1915. The objections of the "dichotomists" can be met by modifying the tree as shown in Fig. 2. If one should accept that the *Brachiella*-branch evolved from the ancestors it shared with the *Lernaeopoda*-branch, rather than being directly descended from the present-day *Lernaeopoda* Blainville, 1822, and that the *Clavella*-branch diverged from the *Charopinus*-branch, one would obtain a phylogenetic tree retaining all the salient features proposed earlier, at the cost of only minor changes.

Fig. 2 also indicates a presumptive relationship between the *Clavella*-branch of Lernaeopodidae and the family Naobranchiidae. A close relationship between these two families is obvious. Both use second maxillae as attachment organs, though they use them in different ways. The latter evolved ribbon-like second maxillae which embrace gill filaments and fuse, in permanent loops, the former use them in conjunction with the bulla (there are some exceptions to this rule, see p. 41) as an anchoring device. The lernaeopodid version is much more versatile and allows this family to exploit a much wider range of habitats on the fish; Naobranchiidae are limited to the gill filaments. The males of Naobranchiidae are morphologically close to those of the *Clavella*-branch of Lernaeopodidae. With the high importance of males in the classification of Lernaeopodidae (cf. Kabata, 1979), one would be inclined to place Naobranchiidae close to the *Clavella*-branch of that family.

One other point of criticism made by G. Fryer (personal communication) against my proposed family tree for Lernaeopodidae is the question of timing the evolution it attempts to represent. I suggested that the *Salmincola*-branch is the most primitive of Lernaeopodidae, evolved in association with the salmonid fishes, which introduced this branch into freshwater habitats. The known history of Salmonidae is short enough to suggest that the *Salmincola*-branch is also a very recent group of species. If true, this would make other branches even more recent and would compress the entire evolution of the family into a span of time so narrow that one would have to postulate a truly galloping rate for lernaeopodid speciation. Again, Fryer's point is valid. To produce an array of more than 250 species and to colonize all the world's oceans, as well as the freshwater habitats of the northern hemisphere, one would expect that more time would be necessary than my proposal appears to allow. This postulate of an inordinately fast rate of evolution for Lernaeopodidae as a whole, is, however, only apparent. In fact, I am concerned about being criticized for holding the opposite views. In a reconstruction of the way in which the genus *Merluccius*, the hake, spread from its original home in the north-east Atlantic to occupy its present range, Ho and I concluded that

Merluccius was in existence about 50 000 000 years ago. At that time it already harboured *Neobrachiella insidiosa* (Heller, 1865), a species of the *Brachiella*-branch of Lernaeopodidae. (A detailed discussion of the origin and dispersal of *Merluccius* and the fate of its copepod parasites is being prepared for publication elsewhere.) It would be reasonable to suppose that the *Salmincola*-branch is phylogenetically older than the *Brachiella*-branch. The ancestors of *Salmincola* Wilson, 1915, must have existed for a long time on some marine fish hosts, before they incorporated Salmonidae into the range of their hosts. The rate of their evolution during that phase was presumably slow. The colonization of freshwater habitats provided a powerful impetus to adaptive radiation and acquisition of non-salmonid freshwater hosts. Taking this impetus into account, it is not too difficult to accept that some 30 species of freshwater lernaeopodids evolved in a powerful, short burst of speciation. The remaining lernaeopodids in their native marine habitats evolved at a more leisurely pace. One might object that the timing suggested here leans to the opposite extreme and ascribes to Lernaeopodidae far too ancient an origin. There is no complete answer to this objection except to point out that *Neobrachiella insidiosa* must have existed more than about 50 000 000 years ago to colonize its present distribution range. The only well documented fossilized parasitic copepod (see p. 17) comes from the lower Cretaceous. Judging from its morphology, marine parasitic copepods have not changed much during the last 50 000 000 years

As mentioned earlier, I recently proposed a revision of the classification of Copepoda based on a concept of their phylogeny (Kabata, 1979). A diagram of this classification is shown in Fig. 3. It involved acceptance of the fact that one cannot discuss the phylogeny of parasitic copepods in isolation from that of their free-living relatives. Parasitism has evolved on several occasions in this abundantly successful order of Crustacea. The scheme was based on two facts (referred to as "major events"), basic to the entire evolution of Copepoda. The first fact was the development of tagmosis in previously metamerically undifferentiated ancestors. The position of the border between the anterior and posterior tagma gave rise to two evolutionary stems: Gymnoplea and Podoplea. First recognized by Giesbrecht (1892), subsequently largely disregarded by the students of parasitic Copepoda, these two stems have an important place in the evolutionary scheme of Copepoda. Gymnoplea comprise those species that have the fifth leg-bearing segment incorporated in the anterior tagma, whereas those with the fifth leg in the posterior tagma are grouped in Podoplea. (The evolution of tagmosis is indicated by S in Fig. 3.) Aberrant Monstrilloida, which have no mouth as adults, must again be left out of these considerations. Two suborders departed from the ancestral type; the newly established Poecilostomatoida and Siphonostomatoida evolved a radically new buccal apparatus. Siphonostomatoida have tubular mouths and stylet-like dentiferous mandibles; Poecilostomatoida have gaping mouths and falcate, pliable mandibles. (The evolution of the new buccal apparatus is indicated by M in Fig. 3.) The evolution of the siphonostome mouth either accompanied or preceded the beginning of parasitism on fishes. It seems to have been a pre-adaptation to it. By a rough count, about 75% of all copepods parasitic on fishes are siphonostomes. Poecilostomes constitute another 20%. Only some 5% belong to Cyclopoida, the suborder which in the new classification is restricted to copepods with the ancestral, gnathostome mouth-parts.

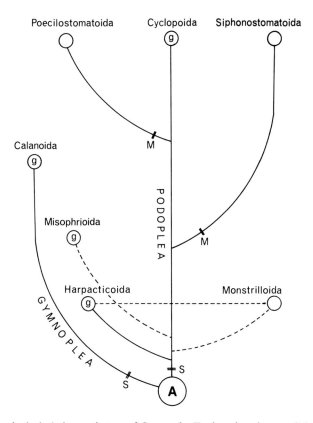

FIG. 3. Hypothetical phylogenetic tree of Copepoda. Explanations in text. (Modified from Kabata, 1979.)

In addition to establishing the two new suborders mentioned above, the proposed classification dismantles Cyclopoida by removing from the old taxon all species with siphonostome and poecilostome mouths. Caligoida are incorporated into Siphonostomatoida and Notodelphyoida into Cyclopoida.

It is worth stressing again that the application of general rules to the problems of classification is difficult. The crux of the matter is the correct choice of appropriate criteria, of suitable morphological characteristics. The clues must be selected anew for each individual problem. A good example of this need for the evaluation of morphological criteria is provided by the fourth leg of the caligid copepods. This leg is in the process of undergoing reduction as the result of the formation of a barrier by the expanded third interpodal bar, a barrier that functionally separates the fourth leg from those anterior to it. The fourth leg has become uniramous in *Caligus* Müller, 1785, and related genera. Its reduction to a vestige consisting of a single segment marks off the genera *Pseudocaligus* Scott, 1901, and *Pseudolepeophtheirus* Markevich, 1940, otherwise indistinguishable from *Caligus* and *Lepeophtheirus* Nordmann, 1832, respectively. In *Caligopsis* Markevich, 1940, the fourth leg had disappeared. The presence or absence of this leg, its structure and size, are used as discriminants at the generic level. *Pseudanuretes* Yamaguti, 1936, is another well defined genus, distinguished by the possession of a unique and prominent flagellum of unknown nature at the base of the second maxilla. This feature sets it off from other caligid genera. *Pseudanuretes*, however, contains species that have vestigial fourth legs, as well as at least one (*Pseudanuretes fortipedis* Kabata, 1965) with full-sized uniramous ones. To be consistent, one would have to split *Pseudanuretes* into two genera. To do so, would be to detract from the diagnostic value of the unique flagellum. By not doing so, one puts in doubt the validity of the fourth leg in this capacity, i.e. the validity of *Pseudocaligus, Pseudolepeopteirus* and *Caligopsis*. A final choice and decision are still awaiting the voice of authority.

A good example of difficulties created by lack of appreciation of what constitutes valid taxonomic criteria is the "superfamily" Dichelesthioidea. This group of about 25 genera was divided by Yamaguti (1963) into five families: Anthosomatidae, Dichelesthiidae, Pseudocycnidae, Eudactylinidae and Catlaphilidae. Yamaguti's arrangement was clearly unsatisfactory. Kabata (1979) reviewed the classification of this ill-defined taxon. The review was based mainly on the type of segmentation, a criterion that had been used in the past to determine familial boundaries among caligiform copepods. The less fusion of segments, the greater the retention of the ancestral metamerism, the more primitive was the copepod judged to be. Four groups were distinguished.

(1) Those with cephalothorax including only maxilliped-bearing segment; other segments clearly demarcated.
(2) Those with three (exceptionally four) segments distinguishable between the cephalothorax and the genital complex.
(3) Those with one or two segments present (when distinguishable) between the cephalothorax and the genital complex.
(4) Those with no free segment between the cephalothorax and the genital complex.

This arrangement resulted in separating genera which on closer scrutiny were found to differ also in the structure of their appendages. The first group coincided with the family Eudactylinidae. The second clearly fell into two incompatible groups, which constituted Dichelesthiidae and a new family Kroyeriidae. The third group also comprised two families: Pseudocycnidae and a new family Hatschekiidae. The last group became another new family, Lernanthropidae. The genus *Prohatschekia* Nunes-Ruivo, 1954, did not fit into any of the above and had to be designated *incertae sedis*. Catlaphilidae have never been adequately described. Their only species, *Catlaphila elongata* Tripathi, 1960, appears to be a poecilostome.

One of the results of this revision was the disappearance of the family Anthosomatidae. The genus *Anthosoma* Leach, 1916, proved to be so close to *Dichelesthium* Hermann, 1804, that their separation was inadvisable. In addition to the type of segmentation, these two genera shared some unique structural features of appendages, and differed from each other only, if spectacularly, because *Anthosoma* had greatly developed elytra, absent from *Dichelesthium*. Those protective structures could be directly related to the type of host–parasite relationship and are clearly secondary characteristics. They certainly do not deserve to be considered as discriminants at familial level.

The fact that the arrangement of genera by segmentation resulted in separating groups with different types of appendages can be interpreted as evidence that segmentation, in this instance, was a valid taxonomic criterion.

The last decade saw one noteworthy sally into a difficult corner of the systematics of parasitic copepods, the problem of host-induced morphological variability. The problem is well enough known to all parasitologists and over the years has caused proliferation of spurious species. Those who deal with the copepod *Lernaea* (Cyclopoida) have always been confronted with the absence of reliable morphological criteria on which to base their specific diagnoses. The circumstances under which the holdfast of the parasite develops within the tissues of the host result in variability of shape which is very hard to categorize and to break up into coherent units. The appendages, on the other hand, appear to be remarkably uniform throughout the genus. In consequence, a lot of uncertainty has been present in the classification of *Lernaea*, uncertainty not yet quite dispelled. Poddubnaya (1973, 1978) attempted to resolve at least some of these difficulties. Her work, published in Russian and including other papers than those quoted here, has so far not been accessible to English-speaking researchers. The papers cited above contain the salient features of her findings. In her 1973 paper, she examined *Lernaea* parasitic on several fish species. On *Cyprinus carpio* she found the "classical" *Lernaea cyprinacea* (L.) which she referred to as "European," as well as another one, morphologically distinguishable from the first, which she named "Asian." The latter is identical with Leigh-Sharpe's (1925b) *Lernaea elegans*. *Aristichthys nobilis* also carried the "Asian" *Lernaea*. *Ctenopharyngodon idella* harboured *Lernaea ctenopharyngodonis* Yin, 1960 (50%), *Lernaea quadrinucifera* Yin, 1960 (25%) and "Asian" *L. cyprinacea* (25%). *Carassius auratus* was parasitized exclusively by the "European" *L. cyprinacea*, while *Ictiolobus bubalis* had only the "Asian" form. The first result of Poddubnaya's work was her recognition of the validity of *Lernaea elegans*, long considered synonymous with *L. cyprinacea*. She then set out to check experimentally the validity of some species. Infection of *Cyprinus carpio* and *Carassius auratus* with the offspring of "Asian" *Lernaea*, i.e. *Lernaea elegans*, produced adult females clearly similar to the maternal species. The same offspring, when developing on *Ctenopharyngodon idella*, produced 16·6% females of *L. elegans* type and 84·4% recognizable as *L. ctenopharyngodonis*. The larvae of *L. ctenopharyngodonis* grew to resemble the maternal individual on *Ctenopharyngodon*, but on *Cyprinus* and *Carassius* they became mainly *L. elegans*; only 8–18% of them retained morphological characteristics of the mother. *Lernaea quadrinucifera* proved to be nothing more than a synonym of *L. ctenopharyngodonis*. The conclusions, elaborated upon later (Poddubnaya, 1978), were as follows. The name *L. cyprinacea* must be restricted to the parasite of *Carassius auratus*. All other cyprinids carry *L. elegans*, one of the forms of which is *L. elegans* morpha *ctenopharyngodonis*. Poddubnaya recognized as valid also two other species: *Lernaea esocina* (Burmeister, 1835) and *Lernaea parasiluri* (Yu, 1938). Her reasons for this recognition are less clear than those on which other conclusions have been based.

The main contribution of Poddubnaya's work is her demonstration that the morphology of the copepod may be substantially influenced by the host and that the influence of the host may be constant and predictable. The experimental transmutation of *L. elegans* into *L. ctenopharyngodonis* and vice versa appears indisputable. No such clear-cut proof has been offered with regard to the differences between *L. cyprinacea* and *L. elegans*. The matter is rendered more complicated by the work of Fratello and Sabatini (1972), who examined the chromosomes of *Lernaea* collected from *Cyprinus carpio*, *Carassius auratus*, *Lepomis gibbosus* and *Gambusia affinis*. The chromosomes were identical ($2n = 16$) in all specimens. They concluded that all these hosts harboured the same species of *Lernaea*, i.e. *L. cyprinacea*. Their work appears to contradict some of Poddubnaya's conclusions. The evidence of chromosomes needs, however, careful evaluation. It is known that some fish species, even belonging to different genera, can have identical sets of chromosomes. Ideally, one should follow Poddubnaya's approach combined with chromosome examination, as promising resolution of the doubts which, in my view, have not yet been resolved. It is to be hoped that these two examples will soon be followed.

Taken in all, the last two decades have brought some fresh initiatives in the field of the phylogeny and classification of Copepoda. It remains to be seen whether they will pass the test of time.

IV. Some Interesting New Forms

New species of copepods parasitic on fishes continued to be described during the last two decades. Some of them are of particular interest, representing new morphological types and indicating that we have not yet seen the entire range of modifications of which copepods are capable. It is not possible to list them all here, but a few examples are worth describing; their choice reflects my preference.

One year before Bocquet and Stock's (1963) review, the late Poul Heegaard (1962) described a new species, *Hyponeo australis*, for which a new family was established (Fig. 4A). Its general habitus resembles Chondracanthidae, but *Hyponeo* is a definite siphonostome. Its affinities within the suborder are not yet clear. The first antennae are multisegmented and relatively unmodified, whereas the second antennae resemble those of Eudactylinidae. *Hyponeo* has typical siphonostome second maxillae; its first maxillae, on the other hand, are unique (Fig. 4B). They bear some slight resemblance to those of Lernaeopodidae. Heegaard knew neither the host nor the locality of his specimens of *Hyponeo*, but the fact that he found them in the collection of the Australian Museum in Sydney suggested that they probably had a southern origin. Very recently, Markevich and Titar (1978) and V. Titar (personal communication) reported the rediscovery of *Hyponeo australis* in the Sea of Okhotsk, on *Paralepis rissoi*. The male of this copepod is still unknown; only its discovery will enable us to determine the position of *Hyponeo* within Siphonostomatoida with any precision. The distance between the two finds of this copepod suggests that its distribution range is probably wider than might have been originally suspected.

Kabata (1968) described a new species, *Shiinoa occlusa* (Fig. 4C) parasitic on the gills of *Scomberomorus commersoni* off the coast of Queensland, Australia. *Shiinoa* has a poecilostome mandible with two falciform blades, so that its subordinal affinities are clear. It is remarkable in having its premandibular segments (pmd in Fig. 4C) greatly enlarged, constituting more than one-third of the total body length. The anterior end is flattened into a roof-like rostrum which, together with very large second antennae, forms a hoop capable of enclosing a piece of the host's tissue. Only one immature female was found. The label read "Habitat: gills." The structure of the second antennae appeared to confirm this statement. Cressey (1975) found more specimens of this species on the same host taken in the Arabian Sea and on *Grammatorcynus bicarinatus* off North Celebes. Two South Atlantic species of *Scomberomorus* (*S. regalis* and *S. maculosus*) were found to be infected with another species of *Shiinoa*, *S. inauris* Cressey, 1975. Cressey collected both female and male specimens of the latter species; the male was attached to the dorsum of the female. All specimens of both species of *Shiinoa* were attached to the nasal laminae of their hosts. The label of my specimen must have been incorrect, or the specimen's habitat unusual for the genus. Later, Cressey (1976) discovered yet another species, *Shiinoa elagata*, on *Elagatus bipinnulatus* in the Carolinas, and on an undetermined species of *Elagatus* in the Gulf of Thailand and between New Hebrides and Solomon Islands. Here we have a genus of at least three species, with very broad geographical distribution, living on fairly easily accessible fish hosts, and yet unknown until a decade ago. In *Shiinoa* we have not only a new type of morphology, but also an example of elusiveness which brings with it the suggestion that many still undiscovered species may be just outside the reach of the investigator.

Cressey and Patterson (1973) share the honour of having discovered the oldest parasitic copepod on record. Fragments of six specimens of a copepod fossil were discovered in the branchial chamber of a fossil fish, *Cladocyclus*

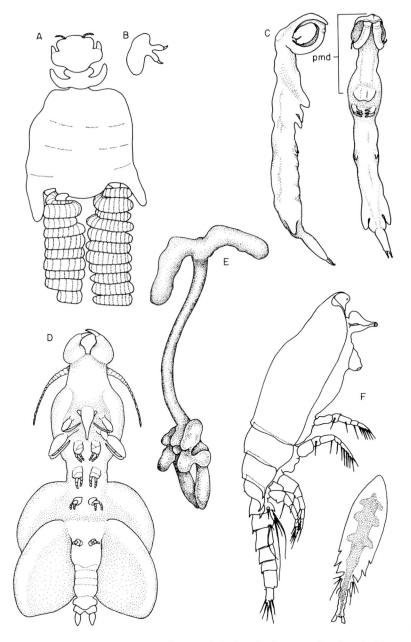

FIG. 4. Some interesting copepods discovered during the last two decades. A, *Hyponeo australis*, female, entire, dorsal view; B, same, first maxilla, lateral view; C, *Shiinoa occlusa*, female, ventral and lateral views (pmd = premandibular segments); D, fossil copepod from lower Cretaceous fish, ventral view; E, *Markevitchielinus anchoratus* female, cephalothorax dorsal, trunk dorsolateral view; F, *Megapontius pleurospinosus*, lateral and outline dorsal views. (A and B from Heegaard, 1962; C from Kabata, 1968; D modified from Cressey and Patterson, 1973; E from Titar, 1975; F from Geptner, 1968.)

gardneri (Ichthyodectidae) from the lower Cretaceous strata in Brazil, dating back 100 000 000 years. One of the specimens was almost complete and it was possible to reconstruct it (Fig. 4D). The specimen shows many close similarities with modern *Dichelesthium*. Its first antenna, consisting of about 20 segments, is its most primitive feature, but the development of two pairs of prominent aliform expansions on the posterior part of the body is an advanced morphological feature, absent from *Dichelesthium*. (One might speculate that *Dichelesthium* is a relative of the Cretaceous form which remained primitive because of its association with Acipenseridae, a family of ancient lineage.) The discovery of this, as yet nameless, fossil copepod greatly extends the time range of parasitic copepods, and of copepods in general. It also suggests that many parasitic copepods have not changed greatly in the course of the last 100 000 000 years.

Titar (1975) described a new genus of Chondracanthidae, *Markevitchielinus*, discovered on a fish (*Hemitripterus villosus*) taken off the South Kurile Islands. Based on a single species, *M. anchoratus* Titar, 1975, this genus is distinguished by a very large, transverse holdfast embedded in the tissue of the fish (Fig. 4E). The entire anterior end of the copepod is buried in the host, as it is in the genus *Strabax* Nordmann, 1864, but unlike its position in the genera of the sub-family Lernentominae, where only premandibular segments are involved in the formation of the holdfast. The development of those relatively gigantic anchoring devices in Chondracanthidae, most of which appear to succeed admirably by using only their second antennae to secure and maintain contact with the host, remains something of a biological puzzle. These copepods do not seem to be in need of such excessive security measures. It is evident that we do not know enough about the biology of Chondracanthidae to explain this enormity.

The last to be mentioned here is the discovery by Geptner (1968) of *Megapontius pleurospinosus*, a free-living copepod. *M. pleurospinosus* Geptner, 1968, deserves mention because it is the second species of the only siphonostome genus not known to live in association with other animals. Taken in the Kurile-Kamchatkan Trench at a depth between 3,860 and 7,100 m, this copepod might be close to the ancestral siphonostomatids. As might be expected, it is a podoplean (Fig. 4F). Its buccal apparatus resembles that of Pennellidae in possessing an elaborate buccal tube; the first and second maxillae are also similar to those of Pennellidae and of several other families of parasitic siphonostomes. Neither the second antenna nor the maxilliped has developed into a prehensile appendage. The general structure is not unlike that of the free-swimming stages of many parasitic copepods. *M. pleurospinosus* must be included in any consideration of the phylogeny of Siphonostomatoida and of the origin of parasitism in this suborder.

V. LIFE CYCLES

The literature of the last two decades contains a fairly large number of reports on the life cycles of parasitic copepods of fishes. Grabda(1963) described the full developmental cycle of *Lernaea cyprinacea*. Lewis (1963) worked out the cycle of *Lepeophtheirus dissimulatus*; Wilkes (1966) that of *Nectobrachia indivisa* Fraser, 1920; and Izawa (1969) that of *Caligus spinosus* Yamaguti, 1939. Shotter (1971) repeated and largely validated earlier work on *Clavella uncinata* (=*C. adunca* (Strøm, 1762)). Kabata (1972) described the cycle of *Caligus clemensi* Parker and Margolis, 1964; Voth (1972) that of *Lepeophtheirus hospitalis* Fraser, 1920; and Zmerzlaya (1972) that of *Ergasilus sieboldi*. Musselius (1967) and Mirzoeva (1972, 1973) published descriptions of the life cycle of *Sinergasilus lieni* Yin, 1949; Kabata and Cousens (1973) that of *Salmincola californiensis* (Dana, 1852); Boxshall (1974a) that of *Lepeophtheirus pectoralis* (Miller, 1776); and Schram (1979) that of *Lernaeenicus sprattae* (Sowerby, 1806). Many publications appeared during the same time containing descriptions of some stages of the life cycle. These are tabulated below.

Name of species	Stage described	Reference
Lernaeenicus sprattae	preadult	Kabata (1963b)
Peniculisa shiinoi Izawa, 1965	copepodid	Izawa (1965)
Cardiodectes sp.	chalimus I–III	Ho (1966)
Sphyrion lumpi (Krøyer, 1845)	nauplius I–II	Jones and Matthews (1968)
Dissonus nudiventris Kabata, 1965	nauplius, copepodid	Anderson and Rossiter (1969)
Cecrops latreilli	chalimus	Grabda (1973)
Sarcotaces pacificus Komai, 1924	nauplius I–V, copepodid	Izawa (1973)
Colobomatus pupa Izawa, 1974	nauplius I–V, copepodid	Izawa (1975)
Bomolochus cuneatus Fraser, 1920	nauplius I	Kabata (1976a)
Chondracanthus gracilis Fraser, 1920	nauplius I	Kabata (1976a)
Ergasilus turgidus Fraser, 1920	nauplius I	Kabata (1976a)
Eudactylina similis	nauplius I	Kabata (1976a)
Haemobaphes diceraus Wilson, 1971	nauplius I, copepodid	Kabata (1976a)
Holobomolochus spinulus (Cressey, 1969)	nauplius I	Kabata (1976a)
Nectobrachia indivisa	copepodid	Kabata (1976a)
Pseudocharopinus dentatus (Wilson, 1912)	nauplius I, copepodid	Kabata (1976a)
Lepeophtheirus hospitalis	nauplius I, copepodid	Lopez (1976)
L. salmonis (Krøyer, 1838)	nauplii	Johanessen (1978)
Dichelesthium oblongum (Abildgaard, 1794)	copepodid	Kabata and Khodorevski (1977)

However, all those studies dealt with the life cycles in a rather narrow and fragmented fashion. At best, they provided comparisons between cycles within relatively small units, genera or families. Since Wilson (1911) drew his well known diagram illustrating five types of life cycles of parasitic copepods, no attempt has been made to take a synoptic view of this subject.

The life cycles of copepods belonging to Gymnoplea, which comprise only the almost exclusively free-living Calanoida, consist of 12 stages other than adult: six nauplii and six copepodids. In contrast, the free-living Podoplea, ancestors of the parasitic copepods, often have only five nauplius stages (although that number may be reduced) and six copepodid stages. Since parasitic copepods descended from podoplean ancestors, they can be expected to have not more, possibly fewer, than five nauplius stages.

Wilson's (1911) diagram implied the existence of trends in the evolution of life cycles, suggesting that the reduction of free-living stages reflected the degree of intimacy achieved within the host–parasite systems. With data accumulated in the course of the last two decades, it is time now to re-examine this implication. It can be most easily done with the aid of another diagram, conceptually based on Wilson's drawing. This diagram (Fig. 5) is based mainly on data obtained within the time under review, but some earlier information has been added. It shows the number of stages, separated from one another by moults, present in the life cycles of various copepods parasitic on fishes, and the mode of life of these stages. Most cycles can be divided into four segments: naupliar, postnaupliar, preadult and adult. With the exception of the adult, these segments commonly consist of more than one stage. I have adopted terminology aimed at elimination of the confusion still existing in labelling individual stages. Nauplii have been numbered I–V and the term metanauplius has been dropped. The postnaupliar stages, beginning with the first copepodid and ending with the last stage preceding the preadult, are clearly affected by the beginning of metamorphosis in some instances and display an advanced stage of sexual maturity in all. These have been named according to their mode of life; those that remain free-swimming and show no extensive organogenetic changes foreshadowing parasitism, are designated by the name copepodid and numbered I–V. If they have become attached and enter the stage of "regressive reconstruction", they are given the name chalimus. There are never more than four chalimus stages and they are numbered accordingly. The next cycle segment, the preadult, is that part during which the copepod either settles definitively on the host and enters a period of metamorphosis, or attains its definitive level of organization, released from the protective restraint of larval semi-permanent attachment.

Remembering that our copepods have descended from the free-living Podoplea, we would expect that the species least modified by their pursuit of a parasitic way of life would have life cycles differing only slightly from those of their ancestors. The evidence at our disposal is, however, far from affirmative with regard to this question. The only known copepods parasitic on fishes and with five naupliar stages in their cycle are *Colobomatus pupa* and *Sarcotaces pacificus*, both greatly modified and endoparasitic poecilostomatoids. Izawa's (1973, 1975) interesting work was successful in obtaining only the first copepodid stages. Although this might denote nothing more than a failure of methodology to assure viability of the larvae long enough to moult into the second (and possibly subsequent) copepodid stages, it might also indicate that the first copepodid is the infective stage, incapable of surviving without the host. The latter alternative is suggested in Fig. 5A. If this alternative is correct, it would seem that the life cycle of Sarcotacidae has been altered by reduction of the free-swimming period to the naupliar segment only.

On the other hand, in the relatively unmodified Ergasilidae (Fig. 5B) only three naupliar stages have remained. Zmerzlaya's (1972) excellent work on the life cycle of *Ergasilus sieboldi* has confirmed that it comprises three nauplii, five copepodids and one "cyclopoid," the last-named being the infective preadult stage. The preadult stage undergoes metamorphosis into adult. In their work on *Sinergasilus lieni*, Musselius (1967) and Mirzoeva (1972, 1973) corroborated Zmerzlaya's findings but differed from Zmerzlaya in failing to recognize the presence of the preadult. It seems to me that the difference was more semantic than substantial. It was determined by the interpretation of the term "stage" and the understanding of what a preadult is.

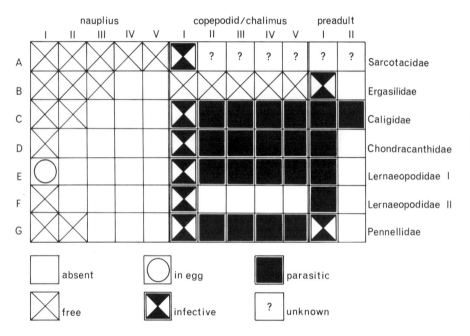

FIG. 5. Types of life cycles of parasitic Copepoda. For explanations, see text.

Further reduction in the number of nauplius stages occurs in Caligidae (Siphonostomatoida), the members of this family having only two nauplii (although Lopez (1976) reported only one in *Lepeophtheirus hospitalis*). The infective stage is the first of the next cycle segment, the copepodid. The remainder of that segment (represented by chalimus stages I–IV) passes in the attached condition, the chalimi being securely fixed to the host by means of the frontal filament, a structure unique to, and prevailing among, Siphonostomatoida. (The reports of some siphonostomatoids developing without the benefit of this larval device require corroboration.) The net effect of this modification of the cycle is a substantial reduction of its free-living part (Fig. 5C). The caligid copepods retain throughout their adult lives their ability to move freely over the surface of the host fish and to change the host individual. This ability begins at the first preadult stage. The life cycles of

Lepeophtheirus dissimulatus (cf. Lewis, 1963) and that of *L. hospitalis* (cf. Voth, 1972) have two preadult stages. Kabata (1972) found only one such stage in the life cycle of *Caligus clemensi*, but this work, being a reconstruction from *in vitro* material, was open to errors. Izawa (1969), on the other hand, failed to find chalimus IV stage in *Caligus spinosus*, though he reported two preadult stages. It is very likely that he and Kabata both missed one stage and that all caligid life cycles have nine stages in addition to the adult. Hewitt (1964) described three "preadult stages" in a caligid cycle. However, his observations on *Lepeophtheirus polyprioni* Hewitt, 1963, did not prove the existence of moults between them. They might be no more than morphological changes occurring with the progress of maturation of one or two stages.

Most of the other known types of life cycles of copepods parasitic on fishes appear to have only one nauplius stage. Heegaard (1947) found that *Acanthochondria cornuta* (Müller, 1776) (Poecilostomatoida) moults from the nauplius stage into a copepodid and that such a copepodid is the infective stage in the cycle (Fig. 5D). His findings need corroboration. They indicate that chondracanthids, after attachment, pass through four other copepodid stages (the name copepodid is used for them because of their lack of the frontal filament). Heegaard maintained this at least in relation to the males of *A. cornuta*. The last copepodid stage, after moulting, undergoes a complicated metamorphosis in a continuous, stepless fashion, which results in attainment of the definitive morphological and biological condition.

The copepods belonging to Lernaeopodidae (Siphonostomatoida) appear to fall into two groups, according to the type of their life cycles. The majority are exemplified by the life cycle of *Salmincola californiensis*, worked out by Kabata and Cousens (1973). In this particular species, the single nauplius remains within the egg (Fig. 5E). The only free-living stage is the infective copepodid which hatches from the egg and settles on the fish, becoming attached by the frontal filament. There follow four chalimus stages, the last of which moults into a preadult. The latter metamorphoses into an adult without further moulting. The life cycles of both sexes are essentially alike, though they differ in timing and in the extent of the final metamorphosis. The other group is represented by *Clavella adunca*, the life cycle of which was worked out anew by Shotter (1971). It might be presumed that most, or all, members of the *Clavella*-branch of Lernaeopodidae have similar life cycles. *Nectobrachia indivisa*, not a member of this branch but sharing with its members the abbreviated condition of the male, shares with *Clavella* also the type of life cycle (Wilkes, 1966) (Fig. 5F). In this type of cycle, the single nauplius is free-swimming. It moults into a copepodid, which is the infective stage. The copepodid becomes attached to the fish by the frontal filament and moults into a stage which Heegaard (1947) referred to as the "pupa." There are no more moults and the pupa becomes an adult by continuous metamorphosis. Heegaard's pupa can, therefore, be equated with the preadult, as defined here.

A life cycle with a pattern unusual among copepods parasitic on fishes is exemplified by Pennellidae (Fig. 5G). Two free-living nauplius stages are followed by the infective copepodid and four chalimus stages. In the latter, this type resembles the life cycle of the non-clavellid Lernaeopodidae (Fig. 5E). The resemblance ends, however, when chalimus IV moults into another free-living stage, an infective preadult, which seeks out the second and definitive host of the cycle. The intermediate host might be the same as the definitive host (e.g. in *Lernaeenicus sprattae*), or it might belong to a completely different group of animals (e.g. cephalopods in *Pennella;* cf. Wierzejski, 1877) or pteropods in *Cardiodectes* (cf. Ho, 1966). It should also be mentioned that in some pennellids the chalimus stage IV appears to be absent. Ho (1966) described the development of *Cardiodectes* sp. in which no chalimus IV was found. Interestingly, most species of *Cardiodectes* have only three pairs of thoracopods, the absence of the fourth pair being clearly linked with the disappearance of the larval stage during which it is normally developed.

This survey would not be complete without reference to Lernaeidae. These important parasites are still something of a puzzle, as far as their life cycles are concerned, in spite of many excellent investigations. They are known to develop through three nauplius stages, five copepodid stages and one preadult (cyclopoid of Grabda, 1963). There are strong indications that at least some lernaeid species have life cycles involving two hosts, whereas others have one-host cycles. Some might even be facultatively two-host parasites, although capable of completing their cycles on one host. Until recently, there was no suggestion that *Lernaea* changed its hosts during development; it settled on its host during the first copepodid stage and remained associated with it for life. Perhaps this was due to the fact that most observations came from controlled environments used for the monoculture of fish. Recent observers of the natural habitat have uncovered evidence suggesting that this is not necessarily an inflexible rule. Fryer (1961) found that in Lake Victoria the adults of *Lernaea cyprinacea sensu lato* lived on *Tilapia*, whereas copepodid larvae infected *Bagrus docmac*. An experimental infection of *Tilapia* with copepodids produced on it adult *L. cyprinacea*. Thurston (1969) found similar situations in Lakes Edward and George. *Lernaea barnimiana* (Hartmann, 1865) lived there as an adult on *Tilapia* and *Haplochromis*, but the copepodids needed *Bagrus* as an intermediate host in the cycle. The lernaean cycle might, therefore, be similar to that of Pennellidae (except for the number of nauplii), or it might differ from it by the absence of the second infective stage and of the intermediate host.

One tends to start a review of life cycles with the expectation that it will show a correspondence between the type of the cycle and the systematic affinity of the species compared, i.e. that related species will prove to have similar cycles. Even though our comparative material is limited, we must conclude that this does not seem to be the case. At least the graphic comparison in Fig. 5 does not support it.

Comparing the life cycles of the three families belonging to the suborder Poecilostomatoida (Ergasilidae, Sarcotacidae and Chondracanthidae), we see at least two different types. The ergasilid type with eight free-swimming stages and an infective preadult is very different from the chondracanthid type with one free-swimming nauplius preceding the infective copepodid. The sarcotacid cycle is not known with certainty, but it appears to fall into a third category, since it comprises five nauplius stages. It must be understood, however, that the situation presented here is grossly oversimplified. Our knowledge is scanty and it is possible that other types of cycles will be discovered in Poecilostomatoida, or even in the three families here compared.

The cycles of siphonostomatoid copepods, at least those presently known, have a reduced number of naupliar stages, but they, too, show marked differences from one taxon to another. The differences are best illustrated by the two types of cycles in a single family, Lernaeopodidae.

The lack of accord between the degree of systematic relationship and the type of life cycle seems to be amply supported by these arguments. One must not lose sight of the fact, however, that this discord might be at least partly due to the imperfection of the classification itself. For example, some suggestions have been made (Gurney, 1934) that Lernaeopodidae should be revised by exclusion of the *Clavella*-branch from the family. Such action would separate groups of species with significantly different life cycles. No such separation, however, would be possible for Lernaeidae with dissimilar or variable cycles.

The other possibility to be explored is the connection between the type of cycle and the type of association between the copepod and host fish in the host–parasite system. Wilson (1911) suggested that the extent of adaptation to the parasitic way of life is paralleled by the reduction in the free-swimming period of the life cycle. He pointed to the only slightly modified Ergasilidae as one end, and the extensively modified Lernaeopodidae as the other end, of the scale. His conclusions were based, however, on data even more scanty than those we have accumulated subsequently. The ergasilids, for all their lack of modification and rather loose type of association with their hosts, have only three naupliar stages, whereas the sarcotacids, greatly modified endoparasitic copepods, have the original ancestral five stages. The latter might also have transferred their infective stage from the preadult to the first copepodid stage. Among Lernaeopodidae, the *Clavella*-branch appears to have eliminated the chalimus segment of the life cycle, in contrast to other branches of the family. Yet it has a free-swimming nauplius, whereas the members of other branches in the family often have intraovular nauplii. In spite of these differences, the definitive morphological adaptations and the mode of life during the adult stage of the female are closely similar. Yet another striking example of incomprehensible differences between types of life cycle is given by the early life history of *Dissonus nudiventris*. Kabata (1979) suggested that Dissonidae should be seen as primitive forerunners of Caligidae, because of the rudimentary state of cephalization. Both families live during their adult stages in association with the surfaces of their hosts, and their modes of life display no appreciable differences. At the same time, while Caligidae have two motile nauplii and a vigorously natatory copepodid, *Dissonus nudiventris* has only one nauplius stage. This larva has become

incapable of locomotion and remains attached to the egg sacs of the maternal
individual by means of enormously elongated balancers. The copepodid of
D. nudiventris is also poorly endowed in locomotory capacity. Thus, the
early part of the life cycle of *Dissonus* is more modified than the corresponding cycle segment of the more "advanced" Caligidae, all their affinities
notwithstanding. (Anderson and Rossiter (1969) reported that they had
failed to find any trace of the frontal filament. This is very unusual for the
siphonostomatoids and could be seen as a primitive character.)

No comment has been made here on metamorphosis, frequently present
during the preadult stage. This process, which is nothing more than the outcome of differential growth, will be discussed below with other aspects of
biology.

Incomplete though our knowledge of life-cycle structure may be, it is much
greater than that of the mechanisms responsible for the sequence of ontogenetic events in the life history of parasitic Copepoda. The literature is
virtually silent on this subject.

Davis (1968), who reviewed the hatching mechanisms of many aquatic
animals, provided a good example of our lack of knowledge by pointing out
contradictory opinions on this mechanism. He compared the views of
Heegaard (1947) with those of Lewis (1963). The former believed that the
nauplii of *Caligus* are expelled from the egg by an increase in intraovular
osmotic pressure; the latter, observing *Lepeophtheirus dissimulatus*, closely
related to *Caligus*, was convinced that hatching is effected by active movements
of the nauplius.

Bird (1968), studying the life cycle of *Lernaea cyprinacea*, noted that
copulation took place at the stage of copepodid V and was followed by
further development, implantation and metamorphosis. The development of
females unable to copulate was arrested at that stage. Clearly, copulation is
a trigger, activating continuing development. What are the pathways involved
in this process? At present, they are beyond even our guesses.

One of the most characteristic features of the life cycle of parasitic copepods
is the metamorphosis, which assumes such striking proportions in Lernaeidae
(Cyclopoida), Chondracanthidae (Poecilostomatoida), Pennellidae and
Lernaeopodidae (Siphonostomatoida). The onset of metamorphosis is
marked by an extraordinary physiological change, which has hitherto
escaped the attention of researchers. Prior to it, these copepods develop by
moulting from stage to stage, in time-honoured crustacean fashion. As soon
as they enter metamorphosis, they begin to grow vigorously without recourse
to that ancestral mechanism. *Lernaeocera branchialis* (L.), for example,
increases in length from about 0.5 to 1.5 mm by a series of moults. Then it
ceases to moult and continues to grow up to the size of 60 mm and to increase
its biomass on a comparatively gigantic scale. Why has its rigid, growth-
limiting cuticle become suddenly pliable and elastic enough to permit growth?
What physiological and anatomical features have made it possible? We have
no answers, even though some of them should not be too difficult to find.
The electron microscope can be used relatively easily to determine at least

the histological changes in the cuticle before and during metamorphosis. Investigation of this curious fact is to be strongly recommended (see p. 55).

The foregoing account shows how imperfect is our understanding of the biology of parasitic copepods. All that we can conjecture from it is that each closely related group of species has evolved a set of intricate relationships with their hosts and that the intricacy extends to the life cycle. We can also assume that it is not only the mode of life of the parasite but also the morphology, biology and ecology of its host that determine the extent and nature of these adaptations. Inasmuch as the adaptations denote departures from the ancestral conditions, they are also to some extent determined by them. There is an observable, though not clearly marked, tendency towards reduction of the free-swimming segment of the cycle correlated with the extent of adaptation to parasitism, but the tendency is not equally pronounced in all large taxa. It appears to be most evident among Siphonostomatoida.

It should be recognized that further studies of life cycles might substantially alter the picture presented above and the conclusions derived from it. There are still entire families of parasitic Copepoda the life cycles of which are completely unknown. For example, the development of Sphyriidae is known only from a brief paper by Jones and Matthews (1968), who established the existence of two free-swimming naupliar stages in *Sphyrion lumpi*. Almost nothing is known of the development of Dichelesthiidae, Hatschekiidae, Lernanthropidae and many other families. With regard to Eudactylinidae, for example, Wilson (1922) published a sketchy description of a nauplius of *Eudactylinodes nigra*. Kabata (1976a) described in detail the same stage in *Eudactylina similis*. Yet knowledge of the ontogeny of Eudactylinidae might prove invaluable to our understanding of the evolution of parasitic Copepoda. They belong, without any doubt, to the most primitive siphonostomatoids (Kabata, 1979), as evident from their morphology. It is difficult to study the parasites of elasmobranchs *in vivo*, but effort expended on such studies would be more than worthwhile. Among other things, they might throw some light on the origin of that peculiarly siphonostomatoid feature, the frontal filament. I strongly recommend this subject to the attention of future researchers.

I also wish to reaffirm my view that the study of life cycles is very important to our understanding of this group of parasites. Although the "retroactive influence" of parasitism might invalidate to some extent the use of life cycles as an aid to systematics, a detailed knowledge of the cycles is going to offer new insights into the biology and evolution of parasitism among Copepoda.

VI. HOST–PARASITE RELATIONSHIPS

The statement that parasitologists cannot agree on the definition of a parasite has by now become a hackneyed ditty. One can, however, perceive the reason for its currency, when one tries to focus on the topic of host–parasite relationship. It is a subject with several facets, each of which occasionally assumes preponderance in its treatment by various authors and can obscure others, sometimes rendering the entire question of parasitism more than ambiguous. The most important and most often approached aspects of this problem are: ecological, physiological and pathological. The first tends to focus on the parasite, the way in which it finds its host and interacts with it (often not different from the way in which free-living organisms locate their target substrates). The second tends to deal with the mechanisms responsible for location and interaction. The third, in contrast, focuses on the host, and the injuries sustained by it due to the presence and activity of the parasite. All three are represented in the literature of the last two decades, though not to the same extent.

A.   HOST LOCATION AND RECOGNITION

The problems of finding the host can be divided into those of long-range location and close-range recognition. The former are only a variety of problems faced by many animals and vary with the ecological conditions within which the host–parasite systems become established. Polyanski (1961) pointed out that formation of the parasite fauna of a marine fish differs essentially from that of a freshwater fish, as suggested by Dogiel (1961). Whereas the freshwater fish initially tends to acquire parasites with direct, one-host cycles, the marine fish first becomes infected with those that include intermediate hosts in their cycles. A freshwater fish, therefore, would include copepods among its earliest parasites, whereas a marine fish would tend to become infected with them later in life. The differences are dictated by the respective sizes of the habitats involved. In more confined freshwater habitats the copepod has a chance of locating the fish when the latter is still unable to feed on intermediate hosts of other parasites. In the sea, these chances are smaller and the fish, in its search for food, actively accumulates those parasites that use food organisms in their cycles. Dogiel's and Polyanski's generalizations contain a tacit implication that the copepod's facility of locating the host is inversely proportional to the distances between them. How does the copepod locate the fish? We are still floundering in assumptions and speculations which assign to various "tropisms" the guiding role in bringing the fish and the copepod together. This train of thought is encouraged by work such as that of Russel (1933), who found that the young fish which congregate in the shadow of floating objects (e.g. logs or medusae) are much more extensively infected with juvenile caligid copepods than those that do not seek such shelters. He inferred that the natural preference for shaded habitats drives these parasites into places where their chances of locating suitable hosts are substantially enhanced. Russel's work, however, is a mere glimpse into the unknown. Much more work is needed on the behaviour of the dispersal stages of parasitic copepods, before anything can be stated on this problem with any probability of validity. At its most basic, work of this kind can be conducted quite simply wherever aquaria are available for holding the host and rearing the larvae of the copepod. Herter's (1927) example shows how much can be achieved by fairly unsophisticated experimentation. The last two decades have not produced this type of work.

One searches in vain for recent signs of interest in this important and intriguing aspect of host–parasite relationships. Only oblique and indirect references can be found, not worthy of serious quotation. It is interesting to note in this context that life-cycle studies produced no convincing evidence of synchronization between the cycles of parasitic copepods and those of their host fishes. Many parasites enhance the chances of locating a new host individual by regulating their reproductive cycles so as to release their infective larvae at the time and place most likely to ensure that they will find it. This has not been demonstrated for Copepoda. Their life cycles appear to be subject to cyclical changes influenced by environmental factors, such as temperature. In this, they obviously acknowledge their free-living ancestry.

Regardless of the means employed, the copepod, having located its host fish, must be able to recognize the host as belonging to the "right" species. There are virtually no data on close-range recognition by parasitic copepods. Schuurmans Stekhoven (1934) attempted to study host recognition by *Lepeophtheirus pectoralis*. He placed a copepod in a vessel containing one host individual (*Platichthys flesus*) and carefully plotted its movements. As a source of information on the topic he intended to study, this work proved to be quite barren. The erratic track of *L. pectoralis*, which had criss-crossed the vessel, passing over, or close to, the fish on more than one occasion, gave no hint of the clues which eventually effected the contact between the host and the parasite. The author concluded that host recognition was tactile and that distant chemoreception was absent. We simply have no idea of the sensory physiology of the copepod and can only surmise the inputs that elicit the attachment reflex. Presumably currents caused by the movements or respiration of the fish are among the factors directing the copepod to it. Presumably also the initial contact is likely in many instances to result in eliciting attachment behaviour, regardless of the compatibility of the host for that copepod. I have seen infective females of *Lernaeocera branchialis* attempting, on contact, to grasp, with their second antennae, the wall of the cavity in a glass slide. Sproston (1942) commented on such juveniles of the same species attaching themselves to empty egg-strings of mature females, and even to artifacts. Here again the mechanisms involved remain completely unknown.

It is not unreasonable to suspect that initial contact, by whatever means effected, is followed by definitive recognition, and acceptance or rejection, of the host. Nothing is known of the mechanisms involved in this process, either, but one can speculate that chemoreception on the part of the copepod might be involved. I have suggested that, at least in Caligidae, chemoreception plays a very important role and that host selection is accomplished with the aid of a newly discovered organ (Kabata, 1974b). This organ, situated in the centre of the anterior margin of the dorsal shield, had been mentioned earlier by Wilson (1905) as a "median sucker." Its real purpose is suggested by its structure. The organ consists of numerous minuscule villiform papillae, packed closely in a well defined field and slightly recessed below the ventral surface of the cephalothorax. In *Caligus clemensi*, this organ (Fig. 6A) is reniform and its villi divided in midventral line by a sulcus. The close-up

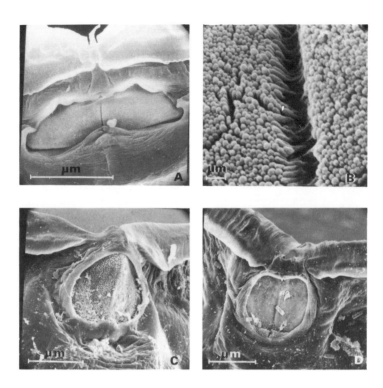

FIG. 6. Sensory (presumably chemoreceptive) organ of Caligidae. Electromicrographs. A, *Caligus clemensi*, entire organ; B, same, detail; C, *Caligus curtus*, entire organ; D, *Lepeophtheirus salmonis*, entire organ. White bars on A, C and D represent 100 μm; on B, 10μm.

view in Fig. 6B, strongly suggests chemoreceptive capability, the surface of the organ resembling that of intestinal endothelium. Similar organs were found in *Caligus curtus* Müller, 1785 (Fig. 6C) and *Lepeophtheirus salmonis* (Fig. 6D). Cressey and Cressey (1979) described them as a "rugose area" in another caligid genus, *Abasia* Wilson, 1908. It is quite probable that all Caligidae possess similar organs and that they are sensory. The organ has not yet been investigated by histological methods, required to support the assertion of its function. Electron microscope studies are needed to confirm it; in particular, its innervation should be studied in great detail. In the meantime, the chemoreceptive function appears at least possible; the position of the organ lends support to this supposition. The point at which it is situated is the first to come in contact with the host, when the latter is located by the copepod. Behaviour experiments on copepods with these organs ablated or masked should be fairly simple to design and can be expected to throw light on the role of this organ. I predict that they will bear out my views on the chemoreceptive function. The question arises, what sensory organs, if any, exist in copepods of other families and suborders, to serve as host selectors? With the electron microscope as a new aid, this problem offers an interesting and rewarding field for investigators.

B. HOST SELECTION

The problems of host location and recognition are further compounded by the fact that copepods do not uniformly infect all individuals within host populations. This implies that either their chances of infecting different host individuals are not equal, or some individuals of the host species are more suitable than others.

The former alternative usually has an ecological basis. For example, Walkey *et al.* (1970) found that the burden of *Thersitina gasterostei* (Pagenstecher, 1861) on sticklebacks, *Gasterosteus aculeatus*, was in part dependent on the salinity of the environment. In water with a salinity of 0.05‰ or less the fish was completely free of the copepod. Since the contact between copepod and fish is maintained in this instance only by the parasite's adhesion to the mucus on the inner surface of the operculum (or some subsidiary surfaces), it is possible that fish infected in areas of higher salinity lose their *Thersitina*, when they move into low-salinity waters. On the other hand, fish that remain in the low-salinity environment are not in danger of infection.

This example leads to the interpretation of the uneven distribution of the copepod on its host population as another aspect of host location, rather than host selection *sensu stricto*. The copepods are able to locate some parts of the population more readily than others. It is well known that fish populations are often spacially segregated by age, or sex, or both. An area ecologically accessible to the copepod might be occupied by fish of a certain age/size group; hence that age/size group is much more likely to acquire the copepod. Should the copepod be permanently fixed and enjoy a long enough life span, subsequent migration of the fish from the "infection zone" would serve to disseminate the parasite. This argument is incomplete, however, until one finds the explanation of the pathways for the eventual return of the copepod to the zone in which it can infect new hosts. Spawning migrations of the host might provide such explanation. This scenario would provide also a possible example of synchronization of life cycles between the host and the copepod.

Alternatively, fish of a certain age/size group might be preferred by the copepod over other groups available in the same habitat. This would naturally lead to uneven distribution of the copepod population. So Fryer (1966), writing about various copepods parasitic on the fishes of African lakes, noted a "marked tendency for a fish which has acquired one parasite to acquire others."

One obvious reason for lack of susceptibility on the part of the fish is the development of immunity. We have no definite evidence that this defence mechanism does operate against copepod parasites, but we cannot reject it out of hand, if only because immunity is a known defence mechanism in fishes, and because of indications that it might affect parasites of other groups.

Another possibility is the physiological state associated with certain age groups and represented by the presence of various hormones or other features which might be found unacceptable to the copepod.

In some instances at least, the fact that the larger and older fish often carry greater numbers of copepods is attributable mainly to the longer period of contact with, and larger attachment surface available for, the parasite. In others, the distribution of parasites appears more random. Boxshall (1974b) found some cod, *Gadus morrhua*, heavily infected with *Clavella adunca* in a population otherwise lightly infected or free of the copepod. His quantitative study (Boxshall, 1974c) pointed to overdispersion, resulting from the adaptation of free-swimming larvae to habitats preferred by the host, as being responsible for the uneven distribution of *Lepeophtheirus pectoralis*. Heegaard (1963) suggested that the patchiness of distribution of *Clavella adunca* might be due to the short life span of its dispersal stages, which are released from the egg sacs in clusters and are likely to result in multiple infections.

Whatever the determinant factor, the commonly occurring increase in copepod burden and prevalence with the size of the fish is well attested. This situation, equally well known among parasites of other groups, has been recorded repeatedly in the literature of the last two decades. To quote but a few examples: Walkey *et al.* (1970) reported an increase in infection of *Gasterosteus aculeatus* by *Thersitina gasterostei* with the size of the fish. Bortone (1971) and Bortone *et al.* (1978) noted a heavier burden of *Ergasilus manicatus* Wilson, 1911, and *Bomolochus concinnus* Wilson, 1911, on larger *Menidia beryllina* and *M. peninsulae*, while Hanek and Fernando (1978c, 1978d) found the same in the association between *Lepomis gibbosus* and *Ambloplitis rupestris*, on the one hand, and three copepod species (*Achtheres ambloplitis* Kellicott, 1880, *Ergasilus caeruleus* Wilson, 1911, and *E. centrarchidarum* Wright, 1882) on the other. Burnett-Herkes (1974) found more *Caligus belones* Krøyer, 1863, *C. coryphaenae* Steenstrum and Lütken, 1861, *C. patulus* Wilson, 1937, *C. productus* Dana, 1852, *C. quadratus* Shiino, 1954, and *Euryphorus nordmanni* Edwards, 1840, on larger *Coryphaena hippurus* than on the smaller ones. Cressey and Collette (1970) distinguished two groups of species in their study of copepod parasites of needlefishes (Belonidae). The first group was designated as "specialized" and defined as "possessing adaptation used as holdfast (and exhibiting higher degree of host specificity)." This group was exemplified by *Lernanthropus* Blainville, 1822, *Caligodes* Heller, 1865 and *Colobomatus* Hesse, 1873. The second group was seen as "generalized" and consisted of "those copepods (Bomolochidae) that are free to wander about (and show less host specificity)". The general premise of increase in infection intensity and prevalence with the size of the host was supported by Cressey and Collette (1970) for the first group but not for the second. The authors attributed the difference between the two groups to the dependence of the specialized copepods on a limited attachment area, which increases with the size of the fish. The generalized copepods, capable of living anywhere on the body surface, suffer from no such restraint. Hence their numbers do not show an increase with the size of the host.

Recent literature contains also records indicating the possible existence of opposite trends in host selection among copepods. Cressey and Collette (1970) themselves recorded *Parabomolochus* (=*Bomolochus*) *bellones* Burmeister, 1835, as occurring more abundantly on smaller fish. Burnett-Herkes (1974), who noted increase in abundance of several *Caligus* species with the size of the host, found that *Caligus bonito* Wilson, 1905, appeared to decrease in abundance; the same was true of *Charopinopsis quaternia* (Wilson, 1936). Dienske (1968) studied the parasite fauna of *Chimaera monstrosa* and found a lernaeopodid copepod, *Vanbenedenia kroeyeri* Malm, 1860, occurring on more than 50% of individuals between 10 and 20 cm long. That prevalence dropped to almost nil by the time the fish reached the length of 25–30 cm. Kabata (1959) examined 104 specimens of *Ch. monstrosa* varying in length from 40 to 100 cm. The fish were of the same population, having been taken in the same net. Only three were infected, carrying eight copepods in all. Joy (1976) found a lowering in the infection rate of *Leiostomus xanthurus* with *Ergasilus lizae* Krøyer, 1863, as the fish became larger. Up to the size of 140 mm, the prevalence was 71% and intensity 4.14; above that size, they dropped to 52% and 3.14 respectively. There are fairly numerous older reports illustrating the same situation.

Noble *et al.* (1963) studied the gill parasites of *Gillichthys mirabilis*, one of the species being *Ergasilus auritus* Markevich, 1940. They found that the number of infected fish increased with length from 70 to 120 mm, but dropped steadily thereafter. A suggestion was made that the reduction was due to the development of immunity.

For the two-host cycle copepods of the family Pennellidae, the difference in the pattern of infection might be produced also by differences in the availability of intermediate hosts. Moser and Taylor (1978) suggested this possibility for *Cardiodectes medusaeus* Wilson, 1908. Kabata (1958) made a similar suggestion in relation to *Lernaeocera branchialis*.

The generalization that ties the increase in prevalence and intensity of copepod infections in a direct manner to the size and age of the host might not, therefore, have universal application. Bortone *et al.* (1978) raised this possibility by pointing out some instances that do not fit it, and by bringing into discussion some instances from outside the copepod field. One is faced with the realization that what might be termed the epidemiology of copepod infections does not really exist. Host selection by the copepod parasite, and its resultant prevalence and distribution on the host population, involve problems of much greater complexity than can be gleaned from the literature surveyed above. Ecological and physiological factors, mainly unknown to us, are brought into play in a different way for different host–parasite systems, creating intricately intertwined patterns of actions and reactions which result in observable but incomprehensible facts. The establishment of a solidly based epidemiology of copepod infections is a challenge facing the new generation of copepod specialists. A scientifically rewarding task, it is also one of considerable practical importance.

C. SITE SELECTION

In common with many other parasites, most parasitic copepods are known to favour specific sites on their host fishes. Depending on the location of these sites (superficial or internal), they have been traditionally classified as ecto- or endoparasites. The overwhelming majority of copepods fall within the former category, but endoparasitic copepods are far from rare. Their true abundance remains unknown at present and tends to be underestimated, simply because of their generally small size, and because their secretion within the tissues of the host renders them difficult to detect. Specific search for endoparasitic copepods often reveals them in unexpectedly large numbers, as witnessed by the work of Richiardi on Colobomatidae in the latter part of the nineteenth century (see Kabata, 1979).

The ectoparasitic copepods, in turn, can be divided into two groups: those that retain the freedom of their movements over the surface of their hosts, at least during the adult segment of the life cycle, and those that select a permanent site and become immovably fixed to it for life. The first group includes copepods such as Caligidae and their allies, as well as Bomolochidae and some related forms. The second group is represented by Lernaeopodidae, Sphyriidae and Chondracanthidae, among many others. Some copepods appear to be physically capable of altering their position on the host, but do it rarely or never (e.g. Pandaridae).

There are, however, copepods which do not fit into the two traditional categories. These copepods, belonging mainly to the family Pennellidae, penetrate deeply into the tissues of their hosts, often reaching vital internal organs, but leave large parts of their bodies exposed to the external environment. Although in the past they were usually classified as ectoparasites, some doubts existed as to the validity of this view. Sundnes (1970) was prompted by these doubts to examine histologically the capsule formed by the host tissues around the embedded part of *Lernaeocera branchialis*. He found it to be of ectodermal origin and treated his findings as confirmation of the ectoparasitic nature of *L. branchialis*. If one were to accept this point of view, one would have to classify as ectoparasitic all those species that are surrounded by host tissues of ectodermal origin. *Sarcotaces* Olsson, 1872, for example, is totally, or almost totally, enclosed in a pouch produced in the wall of the rectum, i.e. a structure of ectodermal origin. It would be altogether far-fetched to describe this almost totally internal parasite as ectoparasitic. This argument applies even more to Colobomatidae, living within the ectodermal lateral line system. Sundnes's criterion cannot be used to demarcate between ecto- and endoparasites.

Kabata (1979) suggested that another category of host–parasitic relationships should be recognized, based on the type of site and mode of attachment. The copepods of this category, Pennellidae in particular, were given the name *mesoparasites*. The main argument against inclusion of these copepods in the ectoparasitic category was the fact that, although a very large part (up to 80%) of each individual protruded from the host, that part was the morphological equivalent of the genital segment only, a relatively minor part of the animal before its metamorphosis. Even a portion of the genital segment was sometimes within the tissues of the host. I find it difficult to classify as ectoparasitic an animal so deeply buried in the host, especially when that animal is intimately associated with the internal organs of the host.

The fact that some copepods are ectoparasitic, while others follow a mesoparasitic or endoparasitic way of life, introduces the topic of site selection. Parasitic copepods are obviously not only host specific but also, to a lesser or greater extent, site specific.

Rudimentary evidence of such specificity has been accumulating in a haphazard fashion for a long time, mainly as a byproduct of descriptions dealing with the occurrence and distribution of copepod parasites on their hosts. It became clear long ago that some copepods occur only on the gills of their hosts, while others preferentially inhabit exposed surfaces of the skin. Other sites of predilection were recorded. With the passing of time, some of these areas were subdivided into recognizable microhabitats, each with a definite degree of suitability for particular copepod species. The trends towards a precise definition of these microhabitats continued during the last two decades, though still as a byproduct of other investigations. A good example of descriptions that produced information on site selection is provided by a number of papers published by Hanek and Fernando (1978a, b, c, d). These authors found that *Ergasilus centrarchidarum* was randomly distributed on the gills of *Lepomis gibbosus*, showing no preference for any special part, but was much more selective on the gills of *Ambloplitis rupestris*, where it preferred the dorsal and ventral sectors of the anterior halves of the hemibranchs to the central sector and to posterior halves of the hemibranchs. No difference between the arches was observed. Fryer (1968) found site preferences in *Ergasilus flaccidus* Fryer, 1965, *E. latus* Fryer, 1959 and *E. kandti* van Douve, 1912, all of which settle on the anterior halves of hemibranchs, always on the ventral half of the first three arches. Kabata and Cousens (1977) studied the attachment of *Salmincola californiensis* to the fry of *Oncorhynchus nerka*, and found that three out of four copepodids initially attached themselves to the areas not covered by scales (fins, fin bases). A disproportionately large number settled on or near the pectoral fins, producing marked concentrations in that small area.

Site selection is undoubtedly determined by a set of morphological and physiological factors, completely unknown at present. Copepods which are semi-mobile on the body surface of the host (and sometimes capable of changing host individuals) tend to congregate in areas that provide shelter and food compatible with their needs, evolved over a long period of formation of the host–parasite systems in which they are partners. For example, *Ergasilus* inhabits mainly the gills of its hosts, presumably having come into contact with them by responding to the respiratory currents of the fish. Its second antennae have become eminently suitable as a grasping apparatus required to maintain a hold on this somewhat uncertain substrate. *Ergasilus*, however, can live also on other substrates, provided they are not covered by scales. Hence, one finds *Ergasilus* at the base of fins or on the fins themselves.

Bomolochus, on the other hand, is rather poorly endowed with a prehensile apparatus and, aided by suction created by the concavity of its ventral surface, lives embedded in mucus on the inner wall of the operculum. It can, therefore, be found only in places protected from currents and capable of producing sufficient quantities of mucus, perhaps stimulated by the pathogenic presence of the parasite. Such places are rare on the surface of the fish (e.g. nasal cavities).

These examples, however, touch only on the observable outcome of the interaction between the copepod and the substrate. They tell us nothing about the mechanisms involved in that interaction. No explanation could be found so far, for example, for the, so-called 'arteriotropism" (Schuurmans Stekhoven, 1936). Some species of *Lernaeocera* Blainville, 1822 and *Haemobaphes* Steenstrup and Lütken, 1861, and virtually all of *Cardiodectes* Wilson, 1917, can unerringly direct their penetration of the host tissues to reach their target site, the cardiac region of the host. Active search for a suitable site has been observed. Kabata and Cousens (1977) described the settling of the copepodid stage of *Salmincola californiensis* on the Pacific salmon hosts. Their observations indicated that the copepodid carefully selects the site at which it excavates the cavity of implantation for the button of its frontal filament. It is not unusual to see the copepodid abandoning a half-excavated cavity and moving to another site. The same authors have found that the site for temporary attachment of the copepodid of this species often does not coincide with the definitive site of permanent attachment of the adult female. In particular, copepodids, which very often attach themselves to the gill filaments, leave them when they have advanced to the preadult stage and search out more secure habitats, especially on juvenile or small hosts. This behaviour suggests that the requirements of the parasite change with its growth and maturation, even when it is destined to become sedentary as an adult.

Very few copepod species are so narrowly specialized that they are unable to survive in a habitat other than their sites of predilection. Although they will preferentially colonize their target sites, once these sites are fully occupied they often spill over to other, less suitable sites. Walkey *et al.* (1970) found this to be true for infections of *Gasterosteus aculeatus* with *Thersitina gasterostei*, but it is equally true of most species, particularly those freely mobile over the surface of the host. This capacity to live in less than optimal habitats has resulted in some astounding feats of survival. T. and A. Scott (1913) described a specimen of *Lernaeocera lusci* (Bassett-Smith, 1846), a parasite normally embedded in a major blood vessel of the host's gills, attached "behind and a little below the base of the pectoral fin". More recently, Boxshall (1974b) found one on the operculum and one on the body surface of the fish. An unidentified species of the same genus was found by van Banning (1974) on the tail (!) of *Trisopterus minutus*.

Fryer's (1966) thought-provoking paper on the "gregarious behaviour" of larval parasitic crustaceans, copepods included, dealt, in addition to the topic of host selection, also with site selection. The fact that in a generally copepod-free host population, one or a few fish might be heavily infected, went hand-in-hand with the fact that these infections were often concentrated in a small part of the habitat available for colonization. For example, *Lernaea hardingi* Fryer, 1956, a copepod capable of developing almost anywhere on the body of its host, *Chrysichthys mabusi*, is none the less usually clustered around the vent of the fish. Many similar examples were reviewed, including a quotation from Gurney (1913), who found that for sticklebacks "it is not uncommon to find one operculum smothered with parasites (*Thersitina gasterostei*) while the other is nearly free from them". Fryer suggested that the mechanism responsible for gregariousness might be similar to that for producing colonies of sessile free-living Crustacea, i.e. a chemical attractant released by the original settler and adsorbed on the substrate. An interesting suggestion, still requiring corroboration. All that can be said at present is that it has not been proved unfounded in the 16 years since it was put forward.

One of the reasons for the clustering of some parasitic copepods can be found in the solitary habits of their hosts. When only isolated individuals of the host species are available for the parasite's settlement, the best strategy might be reinfection of the host harbouring the maternal parasite by the offspring of that parasite. Reduction of the free-swimming sector of the life cycle would assist the larvae in settling close to the maternal individual. Fryer (1966) quoted from his earlier report that larvae "still capable of locomotion also have been found attached alongside firmly anchored adults." In this instance, factors other than the solitary habits of the host were invoked as being responsible. I believe that precisely those habits were responsible for the attachment of six adult specimens of *Vanbenedenia kroeyeri* to a single dorsal spine of *Chimaera monstrosa* (cf. Kabata, 1959). Even more convincing was the occurrence of a cluster of some 50 specimens of *Vanbenedenia chimaerae* (Heegaard, 1962) on the claspers of one male *Chimaera ogilbyi*, described by Heegaard (1962). Kabata (1964) examined 14 females and found that 9 of them bore larval frontal filaments attached to their trunks and even cephalothoraces. Obviously the larvae used them as their initial substrate, later abandoning them for the surface of the host. The life cycle of *Vanbenedenia* is unknown, but Kabata (1964) found that it has four chalimus stages. Some of them were still attached to their frontal filaments glued on to the adult females. As far as is known, the holocephalans are solitary except for their breeding season. It seemed likely that the larvae were offspring of the adults on which they initially settled.

D. ATTACHMENT

Once an appropriate host has been located and contacted, the parasitic copepod faces the task of maintaining that contact for a prolonged period, usually for the remainder of the life cycle. The study of the attachment processes and of their morphological appurtenances is of some relevance to the fascinating problem of the origin of parasitism itself. It is clear that the ability to maintain at least semipermanent contact with the host was a prerequisite for the evolution of parasitism among copepods. It is necessary, therefore, to postulate some morphological preadaptations. They must be found in the prehensile ability of the appendages, the "pre-parasitic" functions of which were directed to feeding, particularly predatory feeding. Other structures aimed at the maintenance of contact between copepod and host have developed subsequently and supported, or superseded, the original devices. One could refer to these attachment structures as primary and secondary respectively. A review of these structures has recently been published (Kabata, 1979), but it would be helpful to recapitulate it briefly at this point.

1. *Primary*

(*i*) The cephalothoracic sucker. This mode of attachment allows free sliding movement over the body of the fish and depends mainly on suction. The ventral concavity of the cephalothorax, sealed around most of its perimeter by a strip of membrane applied to the surface of the host, is partially evacuated of water. A drop in pressure within the enclosed space creates suction, which presses the copepod firmly to the surface of the host. Modified swimming legs close off the posterior margin of the concavity. The hold is enhanced by assistance from the complicated second antennae. This type of attachment occurs in Caligidae and their allies (Siphonostomatoida) and Bomolochidae (Poecilostomatoida). (*ii*) The second antennae are probably the most common, and often the only, appendages serving attachment. They can be grappling (Caligidae, Bomolochidae), stapling (Chondracanthidae), pinching (Pennellidae) or clasping (Ergasilidae). (*iii*) The maxillipeds, appendages designed for prehension, are probably of minor importance in attachment, though among Siphonostomatoida they are often found hooked into the host tissue. It seems that they are at best auxiliary appendages of prehension. (*iv*) Secondary maxillae are used as attachment appendages by Lernaeopodidae. These appendages serve many siphonostomatoids for manipulation of the frontal filament during the process of larval attachment, and during the chalimus segment of the life cycle. In adult female lernaeopodids they have become permanently fused with a special adult attachment device, the bulla. The structure of the bulla has been described in detail by Kabata and Cousens (1972). (Other functions of the bulla will be mentioned below.) (*v*) Some natatory appendages become modified to aid attachment. The first pair of legs in *Nemesis* (Eudactylinidae) form a unique prehensile appendage which, having become a locking device, assists in the maintenance of hold on the host. A similar auxiliary function is performed by a modified third pair of legs in Lernanthropidae.

2. *Secondary*

(*i*) An example of a newly evolved structure is presented by the adhesion pads of Pandaridae (Siphonostomatoida). These pads, with rugose or transversely grooved surfaces, have developed on some previously existing appendages (second antenna, maxilliped, swimming legs), or on the ventral surface of the cephalothorax, and are associated with parasitism on elasmobranch hosts. Possibly the shagreen texture of the skin of these fishes offers a particularly good adhesion surface for these pads. (*ii*) Holdfast structures are simple or complex outgrowths, or sets of outgrowths, arising from the anterior end of the body and anchoring the copepod permanently in the same manner in which a tree is rooted in the ground. They reach luxuriant proportions and cause extensive destruction of the tissues they traverse, although they do not appear to be directly lethal to the fish, even when lodged in its vital organs. (Suggestions that they might have functions other than attachment will be discussed below.) Holdfasts are a characteristic feature of the families Pennellidae and Sphyriidae (Siphonostomatoida) and Lernaeidae (Cyclopoida). They occur also in some poecilostomatoids (Chondracanthidae), where they seem to be somewhat superfluous as a safety device. Kabata (1979) described the formation of holdfasts as the result of the activity of special growth centres (see p. 54). In Pennellidae they appear to develop mainly from the maxilliped segment, an interesting transmutation of a segment bearing prehensile appendages into a generator of a totally different structure, also prehensile in function. Other segments also contribute to the development of holdfasts, but usually to a lesser degree. (*iii*) Frontal filaments, temporary larval attachment structures occurring in most, if not all, siphonostomatoid parasites of fishes. These filaments are produced in the frontal area of the copepod, sometimes as early as the nauplius stage and are discarded by the copepod at the end of the chalimus segment of the life cycle. (*iv*) Bulla. This is an attachment structure unique to Lernaeopodidae, produced in the same region as the frontal filament, extruded with some difficulty through the anterior margin of the cephalothorax, embedded in a specially excavated hollow in the tissue of the fish and fused with the tips of the second maxillae of the parasite.

A				Chondracanthinae
A	**C**			Bomolochidae
A	**H**			Lernentominae
A	**F**	**A**		Cecropidae
A	**F**	**C**		Caligidae
A	**F**	**B**		Lernaeopodidae I
A	**F**	**B**	**H**	Lernaeopodidae II
A	**F**	**A**	**H**	Pennellidae

FIG. 7. Sequence of attachment of parasitic Copepoda to fish hosts. For explanations, see text.

The type of attachment is of paramount importance in determining the character of host–parasite relationship; this fact is strongly underlined by the diagram in Fig. 7. It will be remembered that, in reviewing the types of life cycles, I concluded that little correspondence existed between the modification of the cycle and the extent of adaptation to parasitism. Generally speaking, a similar conclusion emerges from Fig. 7, which shows that the majority of Copepoda parasitic on fishes use more than one type of attachment in the course of their life cycles. Eight different types of "attachment succession" are shown in the diagram. In all instances the second antennae are the initial larval attachment appendages.

The simplest case is exemplified by Chondracanthinae, a subfamily of Chondracanthidae (Poecilostomatoida). Members of this subfamily, having gripped the host tissue with their preadult second antennae, remain attached to it by these appendages for the rest of their lives. The structure of the appendages changes with age and moulting necessitates their repeated re-attachment, but they continue to serve as the main anchoring device. Because of the irritant effect of the copepod's activity (particularly its extrabuccal digestion) on the tissues of the host, these tissues often proliferate around the point of attachment and produce a swelling, covering the cephalothorax and imprisoning the parasite at that point for life. The parasite remains stationary, however, even in the absence of such restraint. The life cycles of Ergasilidae also involve only one type of attachment, by the second antennae. They differ from Chondracanthinae, however, in becoming attached only during the adult segment of the life cycle. Should subsequent research demonstrate that Eudactylinidae and other families of copepods previously placed in Dicheles-thiidae *sensu lato* (cf. Kabata, 1979) have no frontal filament at the copepodid stage, they also could be placed in this category.

Bomolochidae provide an example of a change from larval attachment by the second antennae to adult attachment by means of a cephalothoracic suction cup. Similarly, only one change of attachment occurs in the life cycle of the chondracanthid subfamily Lernentominae and in the cyclopoid family Lernaeidae. In these two instances the larval attachment is replaced by the development of a more or less elaborate holdfast. In Lernentominae, the holdfast is a large structure formed from premandibular segments. It penetrates the tissues of the fish very deeply but the mouth of the parasite remains on the surface of the host. The lernaean holdfast resembles that of Pennellidae. The head of *Lernaea* is entirely embedded in the host.

Cecropidae, a siphonostomatoid family, change their mode of attachment twice. The usual initial attachment is followed by the development of the frontal filament of the chalimus stages. Grabda (1973) confirmed the existence of the frontal filament in *Cecrops latreilli* by her discovery of a chalimus specimen of that species. During the adult segment of the life cycle, the para-site reverts to attachment by means of its second antennae, with or without the assistance of the maxillipeds. Cecropids remain immovably attached during their adult lives, although there appears to be no morphological reason for their immobility. The life cycles of the related family, Pandaridae, are not known with certainty and their possession of the frontal filament is open to question (Cressey, 1967). Should its presence be confirmed, they also would belong in this category. More research on pandarid development is needed.

Caligidae also go through two changes of attachment, the first of them being from the second antennae to the frontal filament. The second change leads to attachment with the aid of the cephalothoracic suction pad, similar to that of Bomolochidae.

The family Lernaeopodidae falls into two categories. Most lernaeopodids change their mode of attachment twice (Lernaeopodidae I in Fig. 7). From the frontal filament of the chalimus segment of the cycle they change to the attachment of a tether-like system of bulla and modified second antennae, the latter undergoing enormous elongation. Some species go one step further (Lernaeopodidae II, Fig. 7). In *Brianella* Wilson, 1915, *Dendrapta* Kabata, 1964, and *Schistobrachia* Kabata, 1964, the bulla remains rudimentary and is replaced by development of the maxillary holdfast. The tips of the maxillae sprout processes, sometimes luxuriant in size and complexity, which take over the function of the bulla as anchoring devices. Thus, in the life cycle of these lernaeopodid genera there are no fewer than four different types of attachment to the host.

Four types of attachment can also be distinguished in the two-host cycles of Pennellidae. Following the chalimus segment of the cycle, they discard the filament and abandon the intermediate host in search of the definitive one. There follows another infective stage (preadult) and another temporary attachment by the second antenna. The definitive attachment is achieved by the development of the cephalothoracic holdfast.

Although no direct correspondence can be found between the attachment type and the extent of parasitic adaptation, some relationship appears to exist between the type of attachment and the definitive size of the adult copepod. One can generalize that smaller parasites have less need of strong attachment than do larger ones. Members of the genus *Hatschekia* Poche, 1902 (Siphonostomatoida), for example, are very small parasites of teleost gills, rarely exceeding a length of 3 mm. Their second antennae are sufficient to ensure their attachment, even without the assistance of the maxillipeds, absent from the females of this genus. Copepods of the genus *Lernanthropus*, of the same order, exceed the size of *Hatschekia* by as much as one order of magnitude and use the second antennae supported by the maxillipeds and the third legs. The small *Eudactylina* van Beneden, 1853, parasitic on the gills of elasmobranchs, requires only the second antennae aided by the maxillipeds to maintain a secure grip on the host. The related but much larger *Nemesis* Risso, 1826, requires more assistance for this purpose; hence its modified first legs, which have become grasping instruments. The largest copepods of all are invariably those with cephalothoracic holdfasts.

Next to nothing is known of the mechanisms initiating and guiding the processes of attachment. Kabata and Cousens (1977) observed a preadult of *Salmincola californiensis* excavating an implantation cavity for its bulla. They were unable to deduce what initiated this "burrowing reflex", but thought that the reflex was deactivated by contact between the excavating appendages and host tissue of appropriate firmness. On fry of *Oncorhynchus nerka*, whose tissues are uniformly soft, the reflex failed to be deactivated and the parasite continued to burrow until it was buried completely within the host. In some instances, young *S. californiensis* burrowed through the visceral cavity and its organs, to emerge on the other side of the fish.

As with so many other aspects of copepod research, much more remains to be done in investigating this entire process than has been done so far.

E.   EFFECTS ON THE HOST

In a book published a decade ago, I reviewed the effects of copepod infections on their fish hosts and divided them into two categories: local and general (Kabata, 1970). Local effects are those limited to the immediate vicinity of the copepod's attachment site and are mainly due to the mechanical influences of its attachment and feeding activities. These effects are as diverse as the types of host–parasite relationships themselves, and their degree of severity varies from almost negligible to fatal. They might be evident on the surface of the copepod, or be concentrated on one or several internal organs. (Some copepod species have become strictly specific to a single organ of the fish, e.g. the eye or the heart.) The severity of the effects also depends on the intensity of infection. The general effects are those which manifest themselves at sites remote from the permanent habitat of the adult parasite. They can be called systemic and they result in overall debility, either difficult to attribute to a specific cause or traceable to a well defined local effect.

Kabata (1976b) suggested that, for practical purposes, the severity of the impact of copepod infections can be roughly predicted from the nature of the host–parasite relationships. Thus, copepods ectoparasitic on teleost hosts tend to be small and their effects relatively harmless, unless the infection is very intensive and concentrated on the gills. The ectoparasites of elasmobranchs are usually larger, immobile and seldom build up intensive infections. Mesoparasitic copepods do not normally occur on elasmobranchs. On teleosts, these copepods usually produce low-intensity infections, but, because they are relatively large, their effects are highly debilitating. (It should be kept in mind that in some instances they appear to affect their hosts far less severely than one would expect from the extent of the mechanical damage they cause.) Endoparasites are almost always quite small (*Philichthys xiphiae* Steenstrup, 1862, being a notable exception) and appear to be relatively harmless. No debilitating effect has been attributed to them so far, though they cause local damage to host tissues.

The effects of parasites on their hosts have always been a popular subject for investigation, and histopathology has been one commonly studied aspect. The last two decades had their share of publications devoted to it. The most important records of those published before 1970 are listed by Kabata (1970). Sundnes (1970) described the histopathological changes produced by *Lernaeocera branchialis* in the tissues of the Atlantic cod, *Gadus morhua*. Joy and Jones (1973) dealt with the pathology of the infection of *Morone chrysops* with *Lernaea cruciata* (Le Sueur, 1824) and Khalifa and Post's (1976) studies involved *Lernaea cyprinacea* in four hosts (*Pimephales pronales, Lepomis cyanellus, Catostomus commersoni* and *Notemigonus crysoleucas*). Natarajan and Nair (1972) described the impact of *Lernaeenicus hemirhamphi* Kirtisinghe, 1933, on *Hemirhamphus xanthopterus*; Logan and Odense (1974) reported on tissue damage caused by *Philorthagoriscus serratus* in *Mola mola*; Grabda (1975) described damage caused by *Haemobaphes diceraus* in *Theragra chalcogramma*. Moser and Taylor (1978) described tissue damage caused by *Cardiodectes medusaeus* in *Stenobrachius leucopsaurus*. A paper by Shields and Goode (1978), in addition to observations on tissue damage, contained a report on experiments in which tissue reactions to a barbed sliver of teflon were studied in a group of *Carassius auratus*, and found to be less extensive than those provoked by the copepod *Lernaea cyprinacea*.

Generally speaking, damage to host tissues appears to be consistent with the type and intensity of the mechanical activity and its attendant influence exerted by the copepods. It shows no special or surprising features. Lüling (1953) pointed out that skin lesions caused by caligoid copepods, regardless of the species, have one common characteristic, the presence of sizeable intra- or subcutaneous blood lacunae. No comparative histopathological study has been conducted on this or related topics during the last two decades.

One of the most interesting types of host–parasite relationships is that between the fish host and the mesoparasitic pennellid copepods. Schuurmans Stekhoven (1936) examined histologically fish hearts invaded by the holdfast of *Lernaeocera branchialis* and found that the copepod does not break through into the lumen of the heart or adjoining blood vessels. Although it does feed on blood, at least in part, it obtains it from the capillary beds proliferating within the capsule which the host's reaction generates around the anterior, embedded part of its body. Sundnes (1970) confirmed these findings. Kabata (1970) suggested that this separation between the parasite and the lumen of the circulatory system is an adaptation without which such host–parasite systems would not be able to exist. Rupture of the wall of a major blood vessel would result in the rapid production of fatal thrombosis. It appears now that such an outcome is far from inevitable. Grabda (1975) found that *Haemobaphes diceraus* penetrates the lumen of bulbus arteriosus, or other parts of the host's heart. She reported damage to cardiac valves. Moser and Taylor (1978) found a similar break in the integrity of the cardiovascular system caused by *Cardiodectes medusaeus* in its host, *Stenobrachius leucopsaurus*. The copepod was reported to be feeding by direct ingestion of blood from the blood vessels. I have examined several species of deep-water myctophid fishes infected with the same copepod and can confirm Moser and Taylor's findings. The well-developed holdfast of *C. medusaeus* and the anterior portion of its cephalo-thorax lodge at the efferent end of the cardiac pump, in the bulbus arteriosus and the ventral aorta (Fig. 8). In this the copepod resembles all others which have the cardiac region as their target site. As a result, the heart pumps blood directly onto the holdfast. The blood must then trickle between the rhizoid branches of the holdfast and be slowed down on its way to the afferent branchial arteries. This slowing of the bloodstream past the buccal orifice of the parasite probably facilitates ingestion of blood by the copepod; I found unbroken erythrocytes in the oesophagus of *C. medusaeus*. The effects of this substantial disruption in the circulation of the fish can only be surmised. Surprisingly, they do not seem to be fatal. The adaptive power of the fish host is truly amazing.

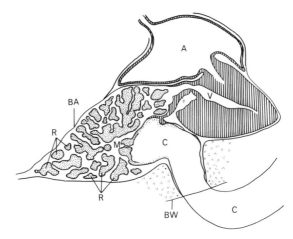

Fig. 8. Section through heart of a myctophid fish with *Cardiodectes medusaeus* embedded in bulbus arteriosus (semidiagrammatic). Abbreviations: A, atrium; BA, bulbus arteriosus; BW, host's body wall; C, copepod; M, mouth; R, rhizoid holdfast processes; V, ventricle.

These observations raise two questions. The first is the nature of the mechanism preventing thrombosis after breach of a major blood vessel. All that we know about the blood-clotting mechanism suggests that the effects of the irruption of the copepod into the vascular system must be nullified by an anticoagulant produced by that parasite. Nothing is known, however, about the biochemistry of this host–parasite interaction or about anticoagulants in Pennellidae. The field is wide open for biochemists, who are now beginning to show signs of interest in parasitic copepods.

The second question is that of the function of the holdfast. It has been suggested (see Kabata, 1970) that, in addition to anchoring the copepod and creating conditions more favourable for the ingestion of blood, the holdfast might be absorptive, secretory, or both. The possibility was first mentioned by Monterosso (1923, 1925, 1926), who found that the so-called "cephalic rhizoids" ("rizoidi cefaliche") of *Peroderma cylindricum* Heller, 1865, contained two vessel-like structures with different reactions to stains. This prompted him to postulate afferent and efferent functions for the two quasi-vessels. The natural next step was the assumption that the dendritic branches of the holdfast were involved in some exchanges between the parasite and the host. Kabata (1970) found similar structures in the branches of the hold-fast of *Phrixocephalus cincinnatus* Wilson, 1908, embedded in the fundus of its host's eye. On the other hand, they exist also in the abdominal brush processes of *Pennella* Oken, 1816, a structure not in contact with the host. They are absent from the holdfast of *Cardiodectes medusaeus*. It appears that our progress in the study of mesoparasitic host–parasite relationships has brought us to another puzzle urgently seeking an answer.

Kabata (1970) listed impact on the growth of the host among the general effects of parasitization with copepods. At its simplest, this involves retardation in the rate of growth of infected individuals. He pointed out the difficulty of evaluating the severity of this effect, subject as it is to modifications by many extraneous factors. One of the atypical results of copepod infection is the promotion of growth, at least in the early stages of infection. Such initial stimulation of growth has been recorded in the infection of *Melanogrammus aeglefinus* by *Lernaeocera obtusa* (=*L. branchialis*) (cf. Kabata, 1958). Two decades later came the report of an apparent promotion of growth of *Stenobrachius leucopsaurus* by a relative of *Lernaeocera*, a pennellid, *Cardiodectes medusaeus* (cf. Moser and Taylor, 1978). The nature of the stimulus is unknown. Kabata (1958) suggested that the effects were due to a compensating over-reaction to the loss of blood, resulting for a time in a polycythaemic condition, higher metabolic rate and more energetic feeding activity. The suggestion is merely speculative and has not been verified experimentally. Such verification is, none the less, not only possible, but reasonably easy to attempt. The stimulating effects of copepod infection on the growth of the fish can only be transitory and must be succeeded by more or less severe retardation.

The effects of copepod infection often include also partial or complete parasitic castration. Kabata (1970) reviewed earlier literature on this topic. Most of the documented instances of castration were associated with meso-parasitic Pennellidae, though *Ergasilus* was also implicated. Recently, Moser and Taylor (1978) added to the list by observing the effect of one more pennellid, *Cardiodectes medusaeus*, on its myctophid host. Infected female fish produced only oocytes, which failed to mature. Multiple infections did not appear to aggravate the effect, neither did the presence of other parasites (nematodes). Since this parasite also appeared to stimulate somatic growth of the host, some relationship between that effect and reproductive retardation appears at least possible.

While more external, macroscopically observable effects of copepod infections have been reasonably well documented and catalogued, far less has been done to determine the impact of the parasites on the metabolism of the host. Mann (1952–1953) showed that *Merlangius merlangus* infected with *Lernaeocera branchialis* consumed less oxygen per unit weight per hour than comparable uninfected fish (0.12 and 0.17 mg respectively). Some depressant influence on metabolism was implied. This decrease in oxygen consumption could be attributed to loss of blood resulting from the feeding of the parasite. Sundnes (1970) related the oxygen consumption of cod, *Gadus morhua*, to the oxygen carrying capacity of its blood. To avoid complicating factors, he experimented with artificially induced secondary anaemia and found that the "observed levels of oxygen consumption and thus the energy metabolism decrease to values below the standard metabolism in the experiments in which haematocrit values were < 10." However, Sundnes did not observe haematocrit values < 20 in cod infected with *Lernaeocera*. The effects of the copepod on the metabolism of the fish could, therefore, not have been severe, if one accepts the validity of these values.

Srinivasachar and Shakuntala (1975) determined the oxygen consumption of the guppy, *Lebistes reticulatus*, infected with *Lernaea hesarangattensis* Srinivasachar and Sundarabai, 1974. The presence of this parasite caused an increase in oxygen consumption of 16.3% in the female and 20.4% in the male fish. When *Lernaea* was removed, the consumption dropped almost to the level of that found in uninfected fish. The parasite *in vitro* consumed oxygen in negligible quantities. The authors assumed, therefore, that the increase in oxygen consumption by infected fish was a "stress reaction" provoked by the parasite.

The above reports present apparently contradictory data. Parasites associated with their hosts in a similar type of relationship seem to provoke diametrically opposite reactions, or, if my interpretation of Sundnes's (1970) data is correct, cause no reaction at all. The contradiction can be resolved only by further well-planned and carefully conducted studies.

F. EFFECTS ON THE PARASITE

Most considerations of host–parasite relationships are slanted in the direction of the effects these relationships have on the host. This attitude is naturally influenced by man's economic interest in the host. It must not, however, obscure the fact that the parasite is also heavily influenced by the host; the relationship is not a one-way traffic.

All animals are influenced by their abiotic, as well as biotic, environment, both of which form the backdrop of evolution and act as guiding forces of considerable magnitude. For parasites, copepods included, the biotic part of the environment dominates the abiotic component, but the general principle of the interplay of species with environment is not thereby altered. One major difference between environment–animal and host–parasite relationships is the fact that in the latter the environment (i.e. the host) has at its disposal defensive mechanisms which can exert a pronounced controlling influence on its inhabitants. The physiological mechanism of immunity is well known in general, though it is rather less well understood with regard to fishes. With reference to its effect on copepod parasites, we know nothing. Some inferences can be indirectly made from observations on differences in the morphology or biology of the copepod, but they are at best conjectural. The entire field is wide open and awaits investigation.

The axiomatic influence of the host on the parasite has been used by systematists to explain morphological differences between members of the same copepod species living on different hosts, or even on different sites on the same host individual. Two references published in the period under review are worth noting in this respect. Cressey and Collette (1970) studied *Lernanthropus belones* Kroyer, 1863, living on the gills of several species of *Strongylura* (Belonidae). They found that the host species could be arranged in a series (*S. timucu–S. marina–S. notata–S. strongylura–S. incisa–S. exilis*) according to the size of the copepod they carried. *Lernanthropus belones* was smallest on *S. timucu* and largest on *S. exilis*. The authors tried to relate this size difference to the size of the host species, but found that the two did not parallel each other. "*S. strongylura*, the smallest of the six host species, does not have the smallest copepods."

Pillai (1970) gave a list of interesting examples to demonstrate that the adaptation to life on the same fish species causes different copepods to develop similar morphological characteristics. The first example showed *Pseudarius jatius*, which harbours three caligid copepod species belonging to different genera (*Caligus arii* Bassett-Smith, 1898, *Lepeophtheirus longipalpus* Bassett-Smith, 1898, and *Hermilius longicornis* Bassett-Smith, 1898). In all three species the apical armature of the first antenna includes highly plumose setae; the armature of the distal exopod segment of the first leg is characterized by the presence of broad membranes unusual for these genera; the endopod of the first leg is larger than is usual and armed with minute prickles; the apron of the third leg is reduced and also armed. Although some of these features (e.g. plumose armature of the first antenna) occur on copepods living on other fishes, they do not occur in the entire combination. The second example featured three species of *Caligus* living on hosts belonging to the family Carangidae (*C. confusus* Pillai, 1961, *C. platurus* Kirtisinghe, 1964, and *C. cordyla* Pillai, 1963). These three species shared seven morphological characteristics, one of which was heavy spinulation of the outer margin of the second segment of the second endopod. This example is rather less convincing. Similar endopod armature occurs in *Caligus* species parasitic on scombrids (*C. bonito, C. productus*) but also infecting fishes of other families. On the other hand, some *Caligus* species living on Scombridae do not have heavily armed second endopods, e.g. *C. pelamydis* Krøyer, 1863. The third example comprises an assemblage of caligid copepods living on *Platax teira*. Here again, four caligids of different genera showed several very characteristic common features. *Heniochophilus branchialis* (Rangnekar, 1953), *Lepeophtheirus anomalus* (Pillai, 1967), *Pseudanuretes schmidti* Rangnekar, 1957, and *Mappates plataxus* Rangnekar, 1958, have large lobes formed from the posterior end of the thoracic zone of the dorsal shield. All of them also have reduced abdomina, so that their uropods appear to project directly from the genital complexes of the thorax. Although the congeners of all these species live on hosts other than *Platax teira*, the coincidence of four copepod species with similar morphological characteristics living on the same host species gives food for thought. In the fourth example, three different caligids (*Tuxophorus wilsoni* Kirtisinghe, 1937, *Lepeophtheirus spinifer* Kirtisinghe, 1957, and *Caligus tylosuri* (Rangnekar, 1956)) parasitic on *Chorinemus lysan* have similar, strongly developed, spiniform fifth legs. One cannot look at all these examples as unequivocal evidence of the host's influence on the parasite. Such influence cannot, however, be ruled out.

As mentioned earlier, one of the ways in which the host affects the parasite is to subject it to the action of its defensive mechanisms. Although the exact nature of these mechanisms remains uncertain, we now have a documented example of the action of such a mechanism. Shields and Goode (1978) described what might be termed self-cure in an infection of *Carassius auratus* with *Lernaea*. They counted 294 already anchored female copepods on day 1 of their experiment. The count on day 6 disclosed only 115 still in place. By day 11 only 15 females (5.1% of the original number) were still attached. There was no doubt that the copepods, well embedded in the tissues of the fish, were removed by rejection, due to the reaction of those tissues. The authors saw this rejection as being apparently due to the integument's response to penetration by the parasite. An earlier paper (Shields and Tidd, 1968) reported a relationship between the success of attachment and the temperature of the environment. Lower temperatures hindered attachment by retarding the progress of the development of the parasite's holdfast processes and by facilitating rejection in this way. At the same time, one might have expected depression of the tissue response at lower temperatures. The picture is not yet quite clear. None-the-less, the existence of defensive mechanisms of the fish has been demonstrated. Kabata (1970) reported on another reaction that eventually led to rejection of the parasite. He wrote: "I found a juvenile female of *Charopinus dubius* Scott, 1900, suspended from the gills on a digitiform outgrowth of soft tissue. The outgrowth was more than 5 mm long (about twice the length of the parasite) and was distinctly pedunculate. It appeared obvious that it would eventually drop off and that the parasite would be removed. I know of no other record of similar occurrence but I assume that the incident observed was not unique."

Taken in all, the relationship between a host organism and a parasite which is dependent on it for its survival is complex and difficult to analyse. This discussion might best be concluded by a quotation from a paper by Noble *et al.* (1963): "Environmental variety generates biological variety which then generates environmental variety. Aquatic chemical and temperature changes that encourage a change in the parasite-mix, a competition for food between two parasites, a mutual exchange of metabolites between host and parasites— all constitute the normal situation we call parasitism. As many of these factors as possible must be studied before we presume to offer final answers to questions on parasite–host relationships."

VII. INTERSPECIFIC RELATIONSHIPS OF PARASITIC COPEPODS

Studies of prevalence and distribution patterns of parasitic copepods have been conducted for a long time. In the course of these studies, researchers from time to time have come upon instances of multiple infections which suggested that the presence of one parasite species at a particular site is not without influence on another species at that site. This influence might be antagonistic or synergistic. Vastly increased sophistication in quantitative biological methods has provided a new and excellent tool for evaluating these quantitative impressions. As in many other aspects of parasitology, however, the parasitic copepods have been left largely outside the sphere of interest. Reliable information on interspecific relationships among parasitic copepods, as well as between copepods and other parasites, is still very scanty.

Two early references to the latter type of relationships have attracted some interest recently. The first was by Wilson (1916), who concluded from his observations that the relationships between glochidia and copepods were antagonistic. The presence of glochidia seemed to preclude, or at least hamper, the settling of copepods on the same site. Much later, Cope (1959) returned to this subject as a result of his studies of fish parasites in Alaska, and concluded that there was no antagonism between *Ergasilus turgidus* and ana-dontan glochidia on the gills of local sticklebacks. Recently, Cloutman (1975), using correlation analysis, corroborated Cope's findings and stated that "there was no indication of antagonism among glochidia, copepods, and other gill parasites."

The "other gill parasites", to which Cloutman referred, included mono-genean flukes. One instance of possible antagonism between a monogenean and a copepod species has caused mild controversy. Leigh-Sharpe (1925a) recorded that *Diclidophora merlangi*, a monogenean, appeared to be mutually ex-clusive with *Clavella adunca*, a copepod, on the gills of *Merlangius merlangus* in British waters. Kabata (1960) supported Leigh-Sharpe's conclusion somewhat tentatively, as *C. adunca*, a common parasite of the gill filaments of *Gadus morhua* and *Melanogrammus aeglefinus*, from which the fluke is absent, avoids the filaments of *Merlangius merlangus*, inhabited by *Diclido-phora*. Smith (1969) disagreed. He pointed out that both copepod and fluke occupy the gill filaments of *Pollachius pollachius* and *Trisopterus luscus*, that *C. adunca* avoids the filaments of specimens of *M. merlangus* even when no flukes are present, and that in those rare instances when it does become attached to them, it is easy to dislodge, unlike specimens attached to the filaments of other fish species. It might be, therefore, that the gill filaments of *M. merlangus* are not a suitable environment for *Clavella adunca*. This example demonstrates how careful one should be in postulating interspecific relationships without exhaustive multifaceted studies.

Cressey (1968) found that three species of *Pandarus* Leach, 1816, living normally on the outer surfaces of *Isurus oxyrhinchus*, are displaced from their usual habitat by *Dinemoura latifolia* (Steenstrup and Lütken, 1861) and/or *D. producta* (Müller, 1785), which compete with *Pandarus* for the substrate. A habitat shift takes place and *Pandarus* moves mainly to sites unusual for it, into the branchial and buccal cavities of the host.

Lewis *et al.* (1969) examined copepods of large Pacific scombrids and concluded from the study of their distribution that *Elytrophora brachyptera* (=*Euryphorus brachypterus* (Gerstaecker, 1855)) appears to compete with *Caligus productus*. Another *Euryphorus*, *E. nordmanni*, was considered by Burnett-Herkes (1974) to have a positive relationship with *C. productus* on the gills of the dolphin, *Coryphaena hippurus*, from the Straits of Florida. Using the chi-square technique, he found that this and four more pairs of parasites were positively related to each other at a 95% significance level. (The members of these pairs were *Caligus coryphaenae*, *C. productus*, *Euryphorus nordmanni*, *Charopinopsis quaternia* (Wilson, 1936), and a species of trematode, in various combinations.) Two pairs (involving *Charopinopsis quaternia*, *Euryphorus nordmanni* and *Caligus productus*) were negatively related at the same level of significance. *Euryphorus nordmanni* is closely related to *E. brachypterus*. We are, then, faced with the possibility that *E. nordmanni* is positively associated with *C. productus*, while *E. brachypterus* is antagonistic to this *Caligus* species. Why this should be so, is not possible to determine with the information now available. Of course, we also might consider that either Lewis *et al.* (1969) or Burnett-Herkes (1974) were incorrect in their conclusions. The third possibility is that the relationship was affected by the environment, one host forcing the two species into antagonism, whereas the other allowed them to develop mutual tolerance.

Cressey and Collette (1970) found competition between species of copepods living on belonid fishes. For example, *Bomolochus bellones* and *Nothobomolochus gibber* (Shiino, 1957) did not occur on the same host individual. Neither did *B. bellones* nor *N. digitatus* Cressey, 1970. *Bomolochus constrictus* Cressey, 1970, and *Bomolochus ensiculus* Cressey, 1970, were rarely found together; when on the same fish, they invariably occupied separate niches. It appeared also that two ergasilid species were rarely present together on the same fish. Antagonistic competitive relationships were suspected.

Ergasilidae and Bomolochidae, on the other hand, did not seem to compete in this manner. Bortone (1971) used the chi-square technique to show that *Bomolochus concinnus* and *Ergasilus manicatus* do not interact negatively. Later (Bortone *et al.*, 1978), he concluded that his earlier findings were based on insufficient evidence, because these two copepod species occur together only rarely, for ecological reasons; the level of interaction is too low to lend itself to definitive analysis. Interestingly enough, the author and his collaborators also found intraspecific avoidance at low levels of intensity.

Rawson (1977) also used the chi-square technique to study relationships between species of copepods, a group of which inhabits the gills of *Mugil cephalus*. The group consisted of *Ergasilus lizae*, *E. versicolor* Wilson, 1911, *E. funduli* Krøyer, 1863, *Bomolochus concinnus*, *Caligus rufimaculatus* Wilson, 1905, *Lernaeenicus longiventris* Wilson, 1917, *Clavella inversa* Wilson, 1913, and *Brachiella oblonga* Valle, 1880. (The last two of these species are of uncertain validity.) He noted their preferential selection of niches, but concluded that, although "competition may be important for different species in the same microhabitat, competitive exclusion did not seem to occur among tested species." His findings tend to corroborate those of Bortone *et al.* (1978).

What has been written above suffices to show that something more than random chance dictates the way in which copepod species occur together on the same host individual. What influences this distribution, it is still impossible to say. Much more work is needed before we can determine whether inter-specific relationships, either positive or negative, are involved.

VIII. BIOLOGY

During the last two decades, investigators concerned with copepods parasitic on fishes have increasingly turned to the study of their subjects as living organisms. Interesting insights have been gained, although we are still only on the threshold of systematic studies of physiology and general biology. The following section will attempt to focus on highlights which might serve to give an overall impression of the state of knowledge of the biology of copepods parasitic on fishes. It will be arranged by topics which bear on particular aspects of life of these parasites.

A. REPRODUCTION

The process of reproduction begins with insemination of the female, the first step of which is the placing of the spermatophores by the male in the genital orifices of the female. Observations of the copulatory behaviour are scanty and the mode of spermatophore transfer poorly known. In Caligidae the copulatory position of the male is on the posterior half of the female, ventral sides of both partners apposed, male second antennae and maxillipeds gripping the female. In Pennellidae the male is attached to the dorsum of the female, its second antennae grasping the partner at about the level of the swimming legs, its genital orifices situated much further posteriorly than those of the female. One assumes that the genital region of the male can be curved and brought to the ventral surface of the female, so that the genital orifices of both sexes will meet. The male of Shiinoidae appears to be attached to the female in the same fashion, though it is not possible to determine whether this is the copulatory position. In Chondracanthidae and Lernaeopodidae the males are dwarves, usually attached near the genital region of the female, with ventral surfaces apposed. Kabata and Cousens (1973) observed the insemina-tion of *Salmincola californiensis*. The male of this species, facing the ventral side of the female, to which it is attached by its maxillipeds, arches the posterior end of its body to bring it close to the space enclosed by the maxil-lipeds. The posterior extremity is then inserted into that space so that the male genital region is apposed to that of the female. This is followed by the extrusion of spermatophores and their introduction into the orifices of the oviducts. It had long been known that, in Lernaeopodidae, these orifices are permanently sealed in adult females by the so-called "brown bodies". These spherical bodies were recognized as remnants of spermatophores and were sometimes found attached to the transparent subpyriform vesicles by narrow and equally transparent ducts. Kabata and Cousens (1973) observed the formation of these curious structures. When the spermatophore is first withdrawn from the body of the male, it appears as in Fig. 9A. Its club-shaped distal end is occupied by three distinct substances. The most proximal of them is a "packet" of spermatozoa, followed by a large droplet of cement substance and another of transparent fluid. Osmotic penetration of water increases the pressure at the distal end of the spermatophore and pushes its contents through the narrow proximal duct of the spermatophore into the oviduct. The cement substance rams the spermatozoa in, acting as a piston, and then permanently seals the oviduct. It assumes its spherical shape (Fig. 9B). The rest of the spermatophore eventually breaks off, leaving the "brown body" as it has been observed for a long time. Rousset *et al.* (1978) restated, on the basis of their observations on the genital apparatus of *Chondracanthus angustatus*, that the further journey of spermatozoa to the seminal vesicle is assisted by the beating of villi ("le mouvement des villosités"). This method of spermatophore transfer explains the tendency towards the evolution of abbreviated males in the *Clavella*-branch of Lernaeopodidae. Such males have no need for elaborate manoeuvres to bring their genital orifices in apposition with those of the female. Their orifices are located close behind the bases of the maxillipeds and become apposed to the oviducts when the male attaches itself to the female.

It has been mentioned above (see p. 26) that insemination might play a role in the development of the female, additional to its obvious reproductive function.

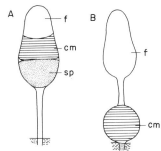

FIG. 9. Spermatophores of *Salmincola californiensis* (diagrammatic). A, spermatophore immediately after introduction into female genital orifice; B, spermatophore after injection of sperm into the female. (From Kabata and Cousens, 1973.) Abbreviations: cm, cement substance; f, fluid; sp, sperm packet.

Morphological studies have demonstrated the structural complexity of the oviduct orifices. It seems clear that this is related to the complexity of the process of oviposition. Most copepods carry their eggs in sacs or strings of various sizes and shapes. The process of their formation and oviposition is, however, still unknown. Heegaard (1963) tried to explain it. Having observed that *Caligus*, which normally has long uniseriate egg strings, produces shapeless multiseriate ones when in captivity and removed from the fish, he concluded that the shaping of the egg strings depends on the stream of water rushing past the flanks of a swimming fish, and pulling out the extruded string into its normal filiform shape. Multiseriate and short egg sacs were due, Heegaard concluded, to the fact that the parasites which produced them lived in places protected from currents. A *Caligus* in the pangs of parturition spurs the fish on to fast swimming by pricking it with the claws of its second antennae and maxillipeds. This explanation is hardly acceptable. It does not take into account the fact that some species or individuals of this genus occupy sheltered niches, where the egg strings must be shaped without the assistance of currents. It would force us into some strange speculation to visualize the formation of the spirally coiled strings characteristic of many siphonostome genera, especially belonging to Pennellidae. Heegaard's attempt notwithstanding, it is true to say that the mechanism of oviposition is still unknown. We should start its investigation with the study of the functional morphology of the appropriate structures, followed by observations of live copepods in the process of depositing eggs.

The differences in the numbers of eggs produced by females of different species have long attracted attention. These numbers range from a few per set of sacs to many hundreds. "The law of the highest possible number of eggs", postulated for parasites in general, obviously has numerous exceptions. Apart from marginal remarks, there has been no attempt to formulate a generalization which would relate the numbers of eggs produced to the biological success of the copepod species. Reddiah (1970), in an abstract of a paper intended for presentation to the II International Congress of Parasitology, tried to review the situation. He saw in the differences between the numbers of eggs a tendency towards increase in the fecundity of parasitic copepods, beginning with some low-fecundity species living on invertebrates (e.g. *Pseudanthessius* spp. with only two or three eggs in a single row) and ending with those producing many hundreds of eggs in multiseriate sacs. From short uniseriate strings, the series moved to long uniseriate ones and then to multiseriate sacs. At the opposite end of the series were the brood chambers of Naobranchiidae. To quote the author: "Parasitic copepods, therefore, demonstrate that in parasitism increased egg production combined with protection of the brood, in the absence of intermediate hosts, is a more favourable adaptation than mere increase in the number of eggs." Reddiah's views on the biological value of the increase in fecundity are much too sweeping. If one judges biological success by the abundance of individuals of any species, one must concede that the genus *Hatschekia* has been very successful. It has radiated into at least 80 species. Some of them are capable of producing, on a single host, populations of more than 100. At the same time, *Hatschekia* is among the least fecund copepods parasitic on fishes. It is also among the smallest. One can suspect the existence of some relationship between the size of the adult female and the number of eggs it can produce. The eggs of small copepods are much larger in proportion to body size than is so with larger copepods. There must be a lower size limit for the copepod egg. The egg biomass produced by the smaller copepods must, therefore, be the outcome of relatively greater reproductive effort. In addition to the size, the type of host-parasite relationship must be taken into account. Probably other factors influencing fecundity exist, though as yet unknown. With regard to Naobranchiidae, their life history is not sufficiently well known to allow us to judge the effects of the brood chambers on their biological success. They are not conspicuously more successful than copepods of similar size and mode of life, which do not possess such chambers. Moreover, it is doubtful whether Naobranchiidae really can be said to have "brood care" in the true sense of that term.

The reproductive process mentioned above, and all other aspects of this process, are subject to environmental influences. The most potent of these is temperature; its influence on reproduction is discussed in the section dealing with temperature (p. 59).

B. GROWTH AND CUTICLE

The earlier section of this paper, dealing with the life cycles of parasitic copepods, made it clear that during the early part of their lives these parasites grow in the same manner as do all other Crustacea, i.e. by stepwise increase in size, through a series of moults. On reaching sexual maturity and the definitive assumption of the parasitic mode of life, they cease to moult. At the same time, their growth becomes much more vigorous. Most copepods parasitic on fishes are larger, some much larger, than their free-living relatives. For some copepods this vigorous growth is accompanied by metamorphosis, as a result of which the adult female parasite becomes completely dissimilar from the early stages.

Kabata (1979) discussed this phenomenon of metamorphosis from the point of view of differential growth. The ultimate difference in shape is due to some parts of the copepod's body growing faster and/or more vigorously than others. The periods of accelerated growth might also differ from one part to another. This is particularly evident in the shape of the early adults. The initial morphological changes, following on attachment to the host, are directed towards the strengthening of the parasite's prehensile and feeding abilities. Consequently, early post-attachment growth is most vigorous in the anterior part of the body associated with these functions. Young adults tend to appear "top-heavy", their posterior halves trailing in development and appearing stunted. When the parasite is securely established, the next phase follows, that of development of reproductive capacity and consequent growth of the posterior part of the body. That part eventually not only reaches the proportions of the anterior part but overtakes it. Gigantism of the reproductive complex is characteristic of many parasitic copepods, especially of mesoparasites. This attainment of the definitive size and shape in two separate phases was referred to as "diphasic growth". In most instances, however, more than two separate growth phases can be identified. Consequently, the existence of several "growth centres" was postulated, each responsible for bringing a particular part of the copepod to its definitive size and shape. The intensity of the activity and the timing of these centres were jointly responsible for the allometric growth of the copepod. More detail is given by Kabata (1979).

As mentioned above, the growth of parasitic copepods undergoes a dramatic change, becoming continuous and stepless. This change has prompted the question (Kabata, 1976b): what has happened to the cuticle? What structural and/or functional modifications occurred at the time of attachment to the host to permit this type of growth? The answer is not yet available. To find it, one must obviously investigate the cuticle before and after that change. No one has yet undertaken this sort of investigation, although it is of great scientific interest. The cuticle of the metamorphosed females of *Pennella elegans* Gnanamuthu, 1957, and *Caligus savala* Gnanamuthu, 1948, was, however, studied in detail by Kannupandi (1975, 1976a, b). He found that the cuticle of *Pennella elegans* consisted of three layers above the epidermis: inner and outer procuticle and epicuticle. The last-named was absent from the part of the cuticle enclosed in the tissues of the host. The epicuticle was partially two-layered, the outer layer being lipid. This layer was absent from the "plumes", i.e. the outgrowths of the abdominal brush. The simple protein of the cuticle contained 13 amino acids, but none contained sulphur or aromatic rings, substances which are involved in hardening. Di- and trityrosine links were found. Kannupandi speculated that they play a part in the stabilization of resilin (a rubber-like compound occurring in arthropod cuticle and skeletal structures) and help the cuticle to remain flexible. Some pore canals were located on the ventral surface of the head. Their function is unknown but the author considered the possibility that they might be absorptive and used in feeding. The cuticle of *Caligus savala* was similar, but its epicuticle was universally two-layered, whereas the procuticle contained calcified and non-calcified strata. Calcification and phenolic tanning were responsible for the hardening of the cuticle. One can only suspect that such hardening does not occur until the copepod reaches its definitive size and little or no further growth takes place. Kannupandi's work is only the first step on the path we must follow to understand the problems of animals that enlarge their size without shedding their exoskeleton.

C. FEEDING

In classifying as parasites copepods that live on or in fishes, one makes an implicit assumption that those copepods live at the expense of the host. Until recently, however, practically nothing was known about the nature of their food and the mode of their feeding. Loose statements about "browsing" or blood-feeding were usually based on superficial and unconfirmed evidence. Some histological studies of the alimentary canal and its contents were conducted, but they were rather perfunctory and fragmentary. The last two decades were not much more enlightening in this respect, though again some studies were undertaken, in some instances quite careful and detailed.

It is obvious that the type of host-derived food depends on the site of the attachment and the entire system of host–parasite relationships. Copepods located on the surface of the host feed on superficial tissues. Shotter (1971) found that *Clavella adunca* fed on mucus. Mature females, removed from the host, when presented with pieces of appropriate tissue, reacted by vigorous movements of their antennae and fed on mucus by "drawing" it in. They did not respond to pieces of scallop. Should such a superficial position be near blood vessels, e.g. on the gills, blood also will be taken. Einszporn (1964, 1965a) studied histologically the alimentary canal of *Ergasilus sieboldi* and found that this copepod fed on gill epithelium, mucus glands, erythrocytes and white blood cells, in that order. The erythrocytes were labelled by ^{59}Fe, so that no doubt could be entertained as to the origin of the material examined in the intestine of the copepod. Deep-seated copepods, such as Lernaeidae, feed on tissue debris, though some of them (e.g. *Afrolernaea*) subsist on a diet of blood (Fryer, 1968). Blood is also ingested by the pennellid copepods. *Cardiodectes*, with the mouth situated directly in the bloodstream of its host, takes in blood as food, as observed by Moser and Taylor (1978) and corroborated by myself (see p. 43). Sundnes (1970) made the point that *Lernaeocera branchialis*, a parasite with a mode of life similar to that of *Cardiodectes*, feeds on whole blood; not only erythrocytes but also serum was ingested.

An obvious difficulty in studying the mode of feeding is the virtual impossibility of direct observation. Most of our ideas on how parasitic copepods feed are derived from our interpretation of the functional role of various appendages involved (presumably) in feeding. Einszporn (1965b) described the feeding of *Ergasilus sieboldi*. She reported it hooking the tissues of the host with its antennae, pushing them with the anterior margin of the cephalothorax and sweeping them towards the mouth by concerted movements of all its swimming legs. She stated that the oral appendages may also tear at the tissues. Clearly, the actions of the latter appendages were not verifiable by direct observation. Fryer (1968) considered the mandibles of *Ergasilus* as being capable of cutting, but did not see them in action. Some of the interpretations of the copepods' mode of feeding are not sufficiently well thought out. One such example is provided by John and Nair (1973), who studied the mouth-parts of *Lernaeenicus hemirhamphi*, a pennellid species. This copepod, like most Pennellidae, has a well developed buccal tube consisting of three rings jointed by thin membranes and apparently capable of being telescoped. The authors suggested the following sequence of events. (1) The buccal tube is pushed in and the rings telescope. (2) The mandibles and the first maxillae are brought into contact with the surface of the host and lacerate it. (3) Blood flows from the lacerations into the buccal tube sealed by the marginal membranes, and is duly ingested. This interpretation visualizes the operation of a free mouth, capable of changing its position in relation to the host's surface. It also endows the mandibles with strength which they probably do not possess. The entire head of *Lernaeenicus* is embedded immovably in a capsule of host tissue. The first maxillae are not directly associated with the procurement of food. They are situated outside the buccal cone and are probably sensory. Fryer (1968), when he referred to *Lernaea*, touched on the secret of how copepods with this type of host–parasite relationships feed. We just do not know how food enters those seemingly impregnable capsules and finds its way into the copepods surrounded by them. Sundnes (1970) did not think that simple sucking could result in whole blood being ingested by *Lernaeocera branchialis*. Kabata (1974a) proposed a novel interpretation of the mode of feeding in Caligidae, based on study of the oral region of that copepod. In this interpretation, the tissue of the host was lacerated by a structure named the strigil, found at the tip of the labium and consisting of two halves, capable of sawing motions. The strigil is armed with a row of small, sharp teeth, in *Caligus clemensi* numbering about 100 in all. Tissue fragments dislodged by the strigil are lifted from the surface of the host and conveyed into the mouth tube by movements of the mandibles in their guiding grooves. Contraction of the intralabral muscles creates a drop in the intrabuccal pressure and facilitates the upward movement of tissue debris. This interpretation can be verified only when instruments such as the new sonoscopes are perfected, allowing us to study feeding processes *in vivo*.

Rigby and Tunnel (1971) suggested that Lernaeopodidae can obtain nourishment from their hosts through the bulla. Cousens (1977) tested this possibility by exposing the bulla of *Salmincola californiensis* to solutions of ^{14}C-labelled amino acids and glucose in salmonid saline. He found that these substances were, indeed, transported across the bulla into the copepod. It is still impossible to say whether this indicates that the bulla plays a significant part in feeding, but such a possibility can no longer be discounted.

An interesting adjunct to the study of feeding is a paper by Lee (1975), who examined the stored lipids of parasitic copepods (*Lepeophtheirus salmonis*, *L. oblitus* Kabata, 1973, *Caligus* sp. and *Clavella perfida* Wilson, 1915). In all species, triglycerides were major storage lipids, as contrasted with the wax esters normally found in free-living copepods. Lee found that the hydrocarbon pattern in parasitic copepods was much closer to that found in the host skin than to that occurring in free-living copepods.

In view of the difficulties in direct observation of feeding activities, little progress can be expected in the near future in our understanding of that process. Much more can be done, however, on the histology, histochemistry and physiology of feeding. Work such as that of Lee (1975) and Cousens (1977) can extend the limits of our knowledge significantly and thus contribute to the understanding of the host–parasite relationships and the attendant practical problems.

D. LOCOMOTION

The great majority of parasitic copepods are immobile as adults. Locomotion is normally associated with the search for food and shelter, or a mate. When the copepod has found its host, the first two of these imperatives have been secured. Mating is sometimes accomplished before attachment to the host. For most parasitic copepods, therefore, locomotion is a vestigial function. It is retained by the free-living dispersal stages of the life cycle. Among siphonostomatoid adults, only Caligidae and their allies can change their position on the host, while among poecilostomatoids only some species of Bomolochidae and related families still have some locomotory ability.

Because locomotion has been tacitly considered unimportant for parasitic copepods, it has never been studied seriously. The movements of nauplii and free-swimming copepodids have been assumed to be effected in a manner similar to that used by free-swimming copepods. They have usually been

dismissed with brief and uninformative statements. Both nauplii and copepodids differ from species to species in the energy with which they move. For
example, while Anderson and Rossiter (1969) referred to the copepodid of
Dissonus nudiventris (Siphonostomatoida) as "lecitotrophic, slow swimming
and demersal", Wilkes (1966) stated that the same stage of *Nectobrachia
indivisa* "darts rapidly about". Shotter (1971) described the copepodids of
Clavella adunca, a species belonging to the same family as *N. indivisa*, as
poor swimmers, propelling themselves through the water by combined
movements of their antennae and swimming legs. Crawling along the bottom,
they employ their maxillae, in addition to the swimming legs. I observed
vigorous swimming of the copepodids of *Salmincola californiensis*, yet
another lernaeopodid which could not be classified as a poor swimmer. The
locomotory activity of the dispersal stages is probably influenced by the mode
of life of the host; it has evolved to maximize the chances of encounter
between the copepod and the fish.

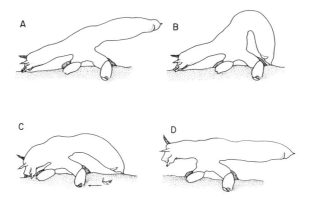

Fig. 10. Four stages in forward movement of male *Salmincola californiensis*. For explanations, see the text. (From Kabata and Cousens, 1973.)

The males of those copepods, which mate on the host, retain enough
freedom of movement to be able to find the female. The females (with the
exceptions mentioned above) remain stationary after final attachment and
the males, having reached sexual maturity, go through a phase of searching,
moving actively over the body of the fish. One such instance was described
by Kabata and Cousens (1973). The mode of progression of the male *Salmincola*, which they observed, can be likened to that of an inchworm. Each
forward movement consists of four steps. (1) The copepod is attached by a
quadruple hold of subchelate second maxillae and maxillipeds; its anterior
end stretches out and is fastened to the substrate by the prehensile second
antennae (Fig. 10A). (2) The trunk is arched so that the uropods, consisting
of several blades each and armed with marginal teeth, are brought right
behind the maxillipeds (Fig. 10B). (3) The second maxillae and the maxillipeds are detached; by simultaneous contraction of the musculature of the

anterior parts of the body and of the trunk muscles the quadruple clamp is shifted forwards (Fig. 10C). The sequence is then repeated (Fig. 10D). It is not known whether this curious mode of locomotion is prevalent among lernaeopodid males: it is certainly impossible among the abbreviated males of the *Clavella*-branch of the family.

Adult Caligidae of both sexes are able to move quite rapidly over the surface of their hosts. They are also able to change the host individual. Kabata and Hewitt (1971) studied the movements of *Caligus clemensi* and *Lepeophtheirus salmonis*, using an inverted microscope and strobe lighting. They showed that only the first two pairs of swimming legs are used for locomotion, which employs a unique type of jet propulsion. In contrast to the prevailing synchronized movements of the swimming appendages in copepods, these two pairs of legs work by opposing each other. There is no usual sequence of effective and recovery strokes. The forward movement of the copepod is effected as follows. (1) The first and second legs move towards each other, pushing out an oblique jet of water posteriorly. (2) The legs move away from each other, allowing the influx of water into the space between them. The second pair, rendered watertight by large membranes and by the downward movement of the endopods, continues to propel the jet of water backwards. The margins of the endopods push against the substrate and momentarily become a fulcrum for forward movement. All these actions are performed under the protection of the dorsal shield, which forms a sucker-like structure (see p. 38). Water is continually drawn under the shield from the anterior direction and leaves its concavity via the posterior sinuses, which are provided with membranes acting as one-way valves. This type of movement is possible even when the host is taken out of water and the jet can no longer be produced. The backward movement of the second leg is sufficient to propel the copepod along the mucus-covered skin.

The most important aspect of the locomotory activities of parasitic copepods is the swimming of the dispersal stages, which is directly responsible for location of the host and survival of the parasite. It is important to study the locomotory mechanisms of copepodids and, in particular, the reasons for the differences in their swimming performance, if we are to understand the evolution of host–parasite systems. This kind of study is reasonably easy to organize. Much can be accomplished without sophisticated equipment and with fairly simple facilities. It is to be hoped that more will be done in this field in the near future.

E. TEMPERATURE TOLERANCE

The influence of ambient temperature on most biological processes is too well known to be debated. It extends to virtually all animals. Parasitic copepods are no exception. In particular, the effects of temperature on their development have been examined almost as often as the development itself. As might be expected, higher water temperatures result in faster development and growth, within certain limits, which differ from species to species and beyond which the temperature becomes retardant or lethal. Some observations on the effect of temperature on the ontogeny of parasitic copepods have been reported during the last two decades. Most of them were simply records of temperature in relation to some ontogenetic events (e.g. Rogers, 1968, 1969; Tedla and Fernando, 1970; Paperna and Zwerner, 1976).

The most interesting experimental study of the effects of temperature and photoperiodicity on the development of parasitic copepods carried out during that period was by Kuperman and Shulman (1972, 1977). These authors used the criterion referred to as the "sum of heat", or degree-days, a combination of temperature and time. They defined it as a factor which "consists of the sum of daily temperatures, taking account of their slight fluctuations". Two series of experiments were performed. The first took place during the winter and early spring season, when the ambient temperature was less than 1°C. The fish (*Esox lucius*) carried on their gills *Ergasilus sieboldi*. They were transferred into aquaria with water temperatures equivalent to those of late spring or autumn (7–9°C), or to those of the summer (10–18°C). In the second series, performed in the autumn, the fish were transferred from an ambience of 7–9°C into aquaria with summer temperatures, or, as controls, temperatures similar to those of the natural habitat. The results of the first series differed from those of the second. During the cold season, even a slight increase in temperature stimulated development of the parasite (gauged by the condition of the gonads), the total effect being dependent on the "sum of heat" applied to the copepod and on the level of temperature before experimental transfer. In January, 160 degree-days caused most of the gonads to develop from stage I to stage II; in February it resulted in attainment of stage III. The second series was carried out in October, and demonstrated that short periods of higher temperature at that time did not stimulate maturation of the gonads. Exposure to temperatures up to 536 degree-days had no effect. Only massive increases (650 degree-days), i.e. exposure to typical summer temperatures of 12–14°C for up to 56 days, caused most of the stage I ovaries to mature to stage II.

The response to the influence of temperature was largely determined by the physiological condition of the copepod. The authors commented on this fact as follows: "The autumn females, preparing for the winter with practically empty ovaries, are unable to start forming egg sacs even under the prolonged influence of a temperature optimal for development. As the spring draws nearer, the process of oogenesis becomes more vigorous, more eggs accumulate in the ovary and the copepod becomes more sensitive to rising temperatures. Consequently, the sum of heat required to prompt the formation of egg sacs becomes progressively smaller." Kuperman and Shulman pointed out, however, that more than the condition of the reproductive organs is involved in the copepod's reaction to the rise in temperature. They suggested "reorganization of the entire organism, its adaptation to a particular season." The biological implications are obvious.

The influence of the photoperiod is also complex. Kuperman and Shulman (1977) established that the formation of egg sacs during spring and early summer is influenced by temperature only. During the late summer and autumn day length is also important. Experiments in October demonstrated that exposure to high temperature had no effect on egg-sac formation. When combined with an extended photoperiod (up to a 12-hour day), however, it stimulated 17–18 % of the females to produce egg sacs. By November, even the combined stimuli of temperature and light failed to prompt gonadal development. The existence of a diapause was postulated.

I believe that Kuperman and Shulman have produced a seminal piece of work which should be a strong cue to future researchers. Once more, relatively simple experiments have given an insight into the biology of a parasitic copepod. Once more the mere fact of asking a question has led to an answer. That answer might be important not only for the development of studies on the basic biology of the copepod, but possibly also for the solution of some practical problems. Biological control of parasitic copepods (in this case of *Ergasilus sieboldi*, a potential threat to aquaculture) is becoming progressively more important because of the difficulties associated with chemical treatment. Manipulation of environmental circumstances under controllable conditions might have merit as a control measure. It cannot be successful, however, without a thorough knowledge of the limits of tolerance, both of the host and of the parasite to be controlled. I believe that much more research of the type described above should be pursued and recommend it particularly to the attention of fishery managers.

F. OSMOLARITY AND SALINITY TOLERANCE

Surprisingly, the problem of the exchanges between parasitic copepods and their external environment has been barely touched by research. The success of any chemical control measure depends on the ability of the controlling agent to penetrate the copepod. It is, therefore, in the interest of all those engaged in aquaculture that this problem be explored and solved. A substantial part of the problem is bound up with the nature of the copepod integument (see p. 54).

It can be assumed *a priori* that copepod–environment exchanges differ from one group of species to another. In particular, extensive differences might be expected to exist between copepods living on freshwater fishes and those parasitizing fishes in a marine environment. One can also expect differences linked with the type of host–parasite system. A high degree of intimacy in the exchanges between the copepod and its fish host is likely to have an impact on the exchanges between the copepod and the external environment.

Lernaea cyprinacea is a typically freshwater species, as are most of its numerous hosts. Some of the fishes it infects are, however, euryhaline. The fact that such hosts might carry *L. cyprinacea* into environments osmotically unusual for the parasite led Shields and Sperber (1974) to investigate "the osmoregulatory relationships including the effects of salinity and the role of the hosts in maintaining the internal concentration of the parasite." Using *Fundulus heteroclitus* as the experimental host, they tested the effects of salinity by transferring the fish into different concentrations of sea water, up to 90%, by different steps and at different rates. They found that *L. cyprinacea*, while attached to its host, was able to maintain its concentration throughout the experiment, with some variations due to its position on the host (depth of penetration) and to the age of the parasite. Highest osmotic values were found in the youngest and the oldest individuals. The authors concluded that "the osmotic concentration of attached *L. cyprinacea* in its normal habitat (freshwater) is not a constant, but rather reflects the interaction of host and parasite physiological factors." Sea water at 20–30% dilution retarded oviposition, and 50% sea water suppressed it completely. Free-swimming and attached pre-metamorphosis females were extremely sensitive to osmotic increases. So were adult females removed from the host. They rapidly became isosmotic with the environment. The authors attributed it to the entry of fluids via the orifices of the alimentary canal. They assumed that adaptation to freshwater habitats caused the evolution of impermeable cuticle. They did not, however, test this presumed impermeability.

In their work with *L. cyprinacea*, Shields and Sperber (1974) referred to the older publication of Panikkar and Sproston (1941). The latter authors studied a marine mesoparasitic copepod of cod (*Gadus morhua*), a pennellid, *Lernaeocera branchialis*, and concluded, like Shields and Sperber, that on excision from the host the copepod becomes rapidly isosmotic with sea water. While on the host, *L. branchialis* was constantly hyperosmotic to the host's blood, but hypo-osmotic to sea water. The role of the host in the maintenance of the copepod's osmotic balance was seen as being of decisive importance. Those findings were questioned recently by Sundnes (1970), who discovered that the intestinal contents of *L. branchialis* are always nearly isosmotic with sea water. The copepod does not appear to have any osmo-regulatory ability. This being so, it must be restricted in its distribution to the part of the host's range in which its osmotic needs can be met. The intimacy of the host–parasite exchanges in the system must also be less than Panikkar and Sproston suggested and the role of the host more limited.

Walkey *et al.* (1970) examined the influence of salinity on the distribution of *Thersitina gasterostei*, parasitic on sticklebacks. No progressive correlation was established. None the less, the absence of the parasite from habitats with 0.5‰ salinity suggested to the authors that salinity might be a limiting factor. Although there are reports that adult *T. gasterostei* can survive in fresh water, it appears that the nauplii are unable to do so and this might determine the inability of *T. gasterostei* to colonize low salinity habitats.

The above example is of interest from the point of view of general parasitology. It is usually accepted that the parasite, to be successful, should be able to tolerate environmental change within broader limits than does its host. Although *T. gasterostei* is capable of parasitizing several species of fishes (Kabata, 1979), its most common important hosts are the sticklebacks, broadly euryhaline species. It would appear that in this instance the parasite became successful in spite of being less resistant than its host to changes in salinity.

Another example of this kind was described by Berger (1970), who studied salinity tolerance of *Lepeophtheirus salmonis*, a parasite of most salmonid fishes. Adult copepods tolerated indefinitely salinity of more than 12‰ but could not survive in lower salinities. Mortality of 100% occurred in about 4 hours, when salinity dropped to 8‰, and water with salinity 4‰ or less proved lethal within 1 hour. The nauplii ceased to swim and died within 5–10 minutes in salinity less than 4‰. At 8‰ they remained insensitive to mechanical stimuli for 1 hour, after which period they became active again; the peak of their activity was attained at 4 hours (only 70% of nauplii survived the initial shock). The peak was followed by gradual depression of activity. Death supervened within 30 hours. These results are rather surprising, since they suggest that the nauplii are more tolerant to changes in salinity, at least above 4‰ level, than are the adult copepods. If this is true, then *L. salmonis* is diametrically different from other parasitic copepods. It is usual for the free-swimming stages to be much more sensitive to environmental influences than the adults.

L. salmonis is a circumpolar species and has been frequently found on all species of *Oncorhynchus* in the Pacific. In British Columbia it has been seen alive on *O. nerka* during the latter's spawning migration as far as 60 miles upstream, in purely fresh waters. This fact poses an interesting question. It is impossible for the salmon to cover that distance upstream within 4 hours. One must assume, therefore, that either Berger's (1970) findings are incorrect and the copepod is much more resistant to salinity changes, or that the physiology of the Pacific *L. salmonis* is different from that of the European population of the species. The latter possibility, if confirmed, would add an interesting dimension to our ideas on the biology of parasitic copepods.

REFERENCES

Anderson, D. T. and Rossiter, G. T. (1969). Hatching and larval development of *Dissonus nudiventris* Kabata (Copepoda, fam. Dissonidae), a gill parasite of the Port Jackson shark. *Proceedings of the Linnaean Society of New South Wales* **93**, 476–481.

Banning, P. van. (1974). Two remarkable infestations by *Lernaeocera* spp. (Copepoda parasitica). *Journal du Conseil International pour l'Exploration de la Mer* **35**, 205–206.

Berger, V. Ya. (1970). [The effect of sea water of different salinity on *Lepeophtheirus salmonis* (Krøyer), an ectoparasite of salmon.] *Parazitologiya* **4**, 136–138. (In Russian.)

Bird, N. T. (1968). Effects of mating on subsequent development of a parasitic copepod. *Journal of Parasitology* **54**, 1194–1196.

Bocquet, C. and Stock, J. H. (1963). Some trends in work on parasitic copepods. *Oceanography and Marine Biology, an Annual Review* **1**, 289–300.

Bortone, S. A. (1971). Ecological aspects of two species of parasitic copepods new to the tidewater silverside, *Menidia beryllina* (Cope). *Journal of the Elisha Mitchell Scientific Society* **87**, 120–123.

Bortone, S. A., Bradley, W. K. and Ogleby, J. L. (1978). The host parasite relationship of two copepod species and two fish species. *Journal of Fish Biology* **13**, 337–350.

Bowman, T. E. (1971). The case of the nonubiquitous telson and the fraudulent furca. *Crustaceana* 21, 165–175.

Boxshall, G. A. (1974a). The developmental stages of *Lepeophtheirus pectoralis* (Müller, 1776) (Copepoda: Caligidae). *Journal of Natural History* 8, 681–700.

Boxshall, G. A. (1974b). Infections with parasitic copepods in North Sea marine fishes. *Journal of the Marine Biological Association of the United Kingdom* 54, 355–372.

Boxshall, G. A. (1974c). The population dynamics of *Lepeophtheirus pectoralis* (Müller): dispersion pattern. *Parasitology* 69, 373–390.

Burnett-Herkes, J. (1974). Parasites of the gills and buccal cavity of the dolphin, *Coryphaena hippurus*, from the Straits of Florida. *Transactions of the American Fisheries Society* 103, 101–106.

Cloutman, D. G. (1975). Parasite community structure of largemouth bass, warmouth and bluegills in Lake Fort Smith, Arkansas. *Transactions of the American Fisheries Society* 104, 277–283.

Cope, O. B. (1959). New parasite records from stickleback and salmon in an Alaska stream. *Transactions of the American Microscopic Society* 78, 157–162.

Cousens, N. B. F. (1977). "Structure and function of the bulla, second maxillae and maxillary ducts of *Salmonicola californiensis* (Dana, 1852) (Copepoda: Lernaeopodidae)." M.Sc. Thesis, University of Victoria, B.C., Canada.

Cressey, R. (1967). Revision of the family Pandaridae (Copepoda: Caligoida). *Proceedings of the United States National Museum* 121, (No. 3570), 1–133.

Cressey, R. (1968). Caligoid copepods parasitic on *Isurus oxyrhinchus*, with an example of habitat shift. *Proceedings of the United States National Museum* 125, (No. 3653), 1–26.

Cressey, R. (1975). A new family of parasitic copepods (Cyclopoida, Shiinoidae). *Crustaceana* 28, 211–219.

Cressey, R. (1976). *Shiinoa elagata*, a new species of parasitic copepod (Cyclopoida) from *Elagatus* (Carangidae). *Proceedings of the Biological Society of Washington* 88, 433–438.

Cressey, R. and Collette, B. B. (1970). Copepods and needlefishes: a study in host-parasite relationships. *Fishery Bulletin* 68, 347–432.

Cressey, R. and Cressey, H. B. (1979). The parasitic copepods of Indo-West Pacific lizard-fishes (Synodontidae). *Smithsonian Contributions to Zoology*, No. 296, 1–71.

Cressey, R. and Patterson, C. (1973). Fossil parasitic copepods from a Lower Cretaceous fish. *Science* 180, 1283–1285.

Davis, C. C. (1968). Mechanisms of hatching in aquatic invertebrate eggs. *Oceanography and Marine Biology, an Annual Review* 6, 325–376.

Dienske, H. (1968). A survey of the metazoan parasites of the rabbit fish, *Chimaera monstrosa* L. (Holocephali). *Netherlands Journal of Sea Research* 4, 32–58.

Dogiel, V. A. (1961). Ecology of the parasites of freshwater fishes. *In* "Parasitology of fishes" (V. A. Dogiel, G. K. Petrushevski and Yu. I. Polyanski, eds.), pp. 1–47. Oliver and Boyd, Edinburgh and London.

Einszporn, T. (1964). [An attempt to determine the mode of feeding of *Ergasilus sieboldi* Nordmann, with the aid of [59]Fe.] *Wiadomosci Parazytologiczne* 10, 530–532. (In Polish, with English summary.)

Einszporn, T. (1965a). Nutrition of *Ergasilus sieboldi* Nordmann. I. Histological structure of the alimentary canal. *Acta Parasitologica Polonica* 13, 71–80.

Einszporn, T. (1965b). Nutrition of *Ergasilus sieboldi* Nordmann. II. The uptake of food and the food material. *Acta Parasitologica Polonica* 13, 373–380.

Ferris, G. F. and Henry, L. M. (1949). The nervous system and the problem of homology in certain Crustacea (Crustacea: Copepoda: Caligidae). *Microentomology* **14**, 114–120.

Fratello, B. and Sabatini, M. A. (1972). Cariologia e sistematica di *Lernaea cyprinacea* L. (Crustacea, Copepoda). *Atti Accademia Nazionale dei Lincei, Memorie Classe di Scienze Fisiche, Matematiche e Naturali* (8) **52**, 209–213.

Fryer, G. (1961). The parasitic Copepoda and Branchiura of the fishes of Lake Victoria and the Victoria Nile. *Proceedings of the Zoological Society of London* **137**, 41–60.

Fryer, G. (1966). Habitat selection and the gregarious behaviour in parasitic Crustacea. *Crustaceana* **10**, 199–209.

Fryer, G. (1968). The parasitic Crustacea of African freshwater fishes: their biology and distribution. *Journal of Zoology, London* **156**, 45–95.

Geptner, M. V. (1968). [Description and functional morphology of *Megapontius pleurospinosus* sp.n. from the Pacific and the position of the genus *Megapontius* in the system of families of the group Siphonostoma (Copepoda, Cyclopoida).] *Zoologicheski Zhurnal* **47**, 1628–1638. (In Russian.)

Giesbrecht, W. (1892). Systematik und Faunistik der pelagischen Copepoden des Golfes von Neapel und der angrenzenden Meeresabschnitte. *Fauna und Flora von Neapel und der angrenzenden Meeresabschnitte* **19**, I–IX and 1–831.

Gotto, R. V. (1979). The association of copepods with marine invertebrates. *Advances in Marine Biology* **16**, 1–109.

Grabda, J. (1963). Life cycle and morphogenesis of *Lernaea cyprinacea* L. *Acta Parasitologica Polonica* **11**, 169–198.

Grabda, J. (1973). Contribution to knowledge of *Cecrops latreillii* Leach, 1816 (Caligoida: Cecropidae), the parasite of the ocean sunfish *Mola mola* (L.). *Acta ichthyologica et piscatoria* **3**, 61–74.

Grabda, J. (1975). Observations on the localization and pathogenicity of *Haemobaphes diceraus* Wilson, 1917 (Copepoda: Lernaeoceridae) in the gills of *Theragra chalcogramma* (Pallas). *Acta ichthyologica et piscatoria* **5**, 13–23.

Gurney, R. (1913). Some notes on the parasitic copepod *Thersitina gasterostei* Pagenstecher. *Annals and Magazine of Natural History*, series 8, **12**, 415–424.

Gurney, R. (1934). Development of certain parasitic copepods of the families Caligidae and Clavellidae. *Proceedings of the Zoological Society of London* **1934**, 177–217.

Hanek, G. and Fernando, C. H. (1978a). Spatial distribution of gill parasites of *Lepomis gibbosus* and *Ambloplites rupestris*. *Canadian Journal of Zoology* **56**, 1235–1240.

Hanek, G. and Fernando, C. H. (1978b). Seasonal dynamics and spatial distribution of *Cleidodiscus stentor* Mueller, 1937, and *Ergasilus centrarchidarum* Wright, 1882, gill parasites of *Ambloplites rupestris*. *Canadian Journal of Zoology* **56**, 1244–1246.

Hanek, G. and Fernando, C. H. (1978c). The role of season, habitat, host age and sex on gill parasites of *Lepomis gibbosus*. *Canadian Journal of Zoology* **56**, 1247–1250.

Hanek, G. and Fernando, C. H. (1978d). The role of season, habitat, host age and sex on gill parasites of *Ambloplites rupestris*. *Canadian Journal of Zoology* **56**, 1251–1253.

Heegaard, P. (1947). Contribution to the phylogeny of the arthropods. Copepoda. *Spolia Zoologica Musei Hauniensis* **8**, 1–236.

Heegaard, P. (1962). Parasitic Copepoda from Australian waters. *Records of the Australian Museum* **25**, 149–233.

Heegaard, P. (1963). Proposed homology in the spawning apparatus of the decapods and the copepods. *Videnskabelige Meddelelser fra Dansk Naturhistorisk Forening* **125**, 311–320.

Herter, K. (1927). Reizphysiologische Untersuchungen an der Karpfenlaus (*Argulus foliaceus* L.). *Zeitschrift für Vergleichende Physiologie* **5**, 283–370.

Hewitt, G. C. (1964). The postchalimus development of *Lepeophtheirus polyprioni* Hewitt, 1963 (Copepoda: Caligidae). *Transactions of the Royal Society of New Zealand, Zoology* **4**, 157–159.

Ho, Ju-shey. (1966). Larval stages of *Cardiodectes* sp. (Caligoida: Lernaeoceriformes), a copepod parasitic on fishes. *Bulletin of Marine Science* **16**, 159–199.

Ho, Ju-shey. (1978). On the origin of Chondracanthidae, a family of Copepoda parasitic on marine fish. *Proceedings of the Fourth International Congress of Parasitology*, Section A, 69.

Izawa, K. (1965). A new parasitic copepod of the genus *Peniculisa* Wilson from Seto, Wakayama Prefecture, Japan. *Reports of Faculty of Fisheries, Prefectural University of Mie* **5**, 365–373.

Izawa, K. (1969). Life history of *Caligus spinosus* Yamaguti, 1939, obtained from cultured yellow tail, *Seriola quinqueradiata* T. & S. (Crustacea: Caligoida). *Reports of Faculty of Fisheries, Prefectural University of Mie* **6**, 127–157.

Izawa, K. (1973). On the development of parasitic Copepoda: I. *Sarcotaces pacificus* Komai (Cyclopoida: Philichthyidae). *Publications of Seto Marine Laboratory* **21**, 77–86.

Izawa, K. (1975). On the development of parasitic Copepoda: II. *Colobomatus pupa* Izawa (Cyclopoida: Philichthyidae). *Publications of Seto Marine Laboratory* **22**, 147–155.

Johanessen, A. (1978). Early stages of *Lepeophtheirus salmonis* (Copepoda: Caligidae). *Sarsia* **63**, 109–176.

John, S. E. and Nair, N. B. (1973). Structure of the mouth tube and method of feeding in *Lernaeenicus hemirhamphi* Kirtisinghe, a parasitic copepod. *Zoologischer Anzeiger* **190**, 35–40.

Jones, D. H. and Matthews, B. L. (1968). On the development of *Sphyrion lumpi* (Krøyer). *Crustaceana*, Supplement 1, 177–185.

Joy, J. E. (1976). Gill parasites of the spot *Leiostomus xanthurus* from Clear Lake, Texas. *Transactions of the American Microscopical Society* **95**, 63–68.

Joy, J. E. and Jones, L. P. (1973). Observations on the inflammatory response within the dermis of a white bass, *Morone chrysops* (Rafinesque) infested with *Lernaea cruciata* (Copepoda: Caligidae). *Journal of Fish Biology* **5**, 21–23.

Kabata, Z. (1958). *Lernaeocera obtusa* n. sp.; its biology and its effects on the haddock. *Marine Research, Department of Agriculture and Fisheries for Scotland* 1958 (No. 3), 1–26.

Kabata, Z. (1959). *Vanbenedenia kroyeri* (Copepoda parasitica): taxonomic review and other notes. *Annals and Magazine of Natural History*, series 13, **2**, 731–735.

Kabata, Z. (1960). Observations on *Clavella* (Copepoda) parasitic on some British gadoids. *Crustaceana* **1**, 342–352.

Kabata, Z. (1963a). *Clavella* (Copepoda) parasitic on British Gadidae: one species or several? *Crustaceana* **5**, 64–74.

Kabata, Z. (1963b). The free-swimming stage of *Lernaeenicus* (Copepoda parasitica). *Crustaceana* **5**, 181–187.

Kabata, Z. (1964). On the adult and juvenile stages of *Vanbenedenia chimaerae* (Heegaard, 1962) (Copepoda: Lernaeopodidae) from Australian waters. *Proceedings of the Linnaean Society of New South Wales* **89**, 254–267.

Kabata, Z. (1966). Comments on the phylogeny and zoogeography of Lernaeopod-idae (Crustacea: Copepoda). *Proceedings of the First International Congress of Parasitology* **2**, 1082–1083.

Kabata, Z. (1968). Copepoda parasitic on Australian fishes. VII. *Shiinoa occlusa* gen. et sp. nov. *Journal of Natural History* **2**, 497–504.

Kabata, Z. (1970). Crustacea as enemies of fishes. *In* "Diseases of fishes" (S. F. Snieszko and H. R. Axelrod, eds), Book 1. TFH Publishers, Jersey City.

Kabata, Z. (1972). Developmental stages of *Caligus clemensi* (Copepoda: Caligidae). *Journal of the Fisheries Research Board of Canada* **27**, 1571–1593.

Kabata, Z. (1974a). Mouth and mode of feeding of Caligidae (Copepoda), parasites of fishes, as determined by light and scanning electron microscopy. *Journal of the Fisheries Research Board of Canada* **31**, 1583–1588.

Kabata, Z. (1974b). Two new features in the morphology of Caligidae (Copepoda). *Proceedings of the Third International Congress of Parasitology* **3**, 1635–1636.

Kabata, Z. (1976a). Early stages of some copepods (Crustacea) parasitic on marine fishes of British Columbia. *Journal of the Fisheries Research Board of Canada* **33**, 2507–2525.

Kabata, Z. (1976b). A rational look at parasitic Copepoda and Branchiura. *In* "Wildlife Diseases" (L. A. Page, ed.), pp. 175–181. Plenum Press, New York and London.

Kabata, Z. (1979). "Parasitic Copepoda of British Fishes". Ray Society, London.

Kabata, Z. and Cousens, B. (1972). The structure of the attachment organ of Lernaeopodidae (Crustacea: Copepoda). *Journal of the Fisheries Research Board of Canada* **29**, 1015–1023.

Kabata, Z. and Cousens, B. (1973). Life cycle of *Salmincola californiensis* (Dana, 1852) (Copepoda: Lernaeopodidae). *Journal of the Fisheries Research Board of Canada* **30**, 881–903.

Kabata, Z. and Cousens, B. (1977). Host-parasite relationships between sockeye salmon, *Oncorhynchus nerka*, and *Salmincola californiensis* (Copepoda: Lernaeopodidae). *Journal of the Fisheries Research Board of Canada* **34**, 191–202.

Kabata, Z. and Gusev, A. V. (1966). Parasitic Copepoda of fishes from the collection of the Zoological Institute in Leningrad. *Journal of the Linnaean Society of London, Zoology* **46**, 155–207.

Kabata, Z. and Hewitt, G. C. (1971). Locomotory mechanisms in Caligidae (Crustacea: Copepoda). *Journal of the Fisheries Research Board of Canada* **28**, 1143–1151.

Kabata, Z. and Khodorevski, O. A. (1977). [The copepodid stage of *Dichelesthium oblongum* (Abildgaard, 1974), a parasitic copepod of acipenserid fishes.] *Parazitologiya* **11**, 236–240. (In Russian.)

Kannupandi, T. (1975). Cuticular adaptation in a copepod parasite *Pennella elegans*. *Current Science (Bangalore)* **44**, 740–741.

Kannupandi, T. (1976a). Cuticular adaptations in two parasitic copepods in relation to their mode of life. *Journal of Experimental Marine Biology and Ecology* **22**, 235–248.

Kannupandi, T. (1976b). Occurrence of resilin and its significance in the cuticle of *Pennella elegans*, a copepod parasite. *Acta Histochemica* **56**, 73–79.

Khalifa, K. A. and Post, G. (1976). Histopathological effect of *Lernaea cyprinacea* (a copepod parasite) on fish. *Progressive Fish-culturist* **38**, 110–113.

Kuperman, B. I. and Shulman, R. E. (1972). [An attempt at experimental investigation of the effect of temperature on some parasites of the pike.] *Vestnik Leningradskogo Universiteta, Biologiya* **3**, 5–15. (In Russian.)

68   Z. KABATA

Kuperman, B. I. and Shulman, R. E. (1977). [On the influence of some abiotic factors on the development of *Ergasilus sieboldi* (Crustacea, Copepoda).] *Parazitologiya* **11**, 117–121. (In Russian.)

Lee, R. F. (1975). Lipids of parasitic copepods associated with marine fish. *Comparative Biochemistry and Physiology, Series B* **52**, 363–364.

Leigh-Sharpe, W. H. (1925a). A revision of the British species of *Clavella* (Crustacea, Copepoda), with a diagnosis of new species: *C. devastatrix* and *C. invicta*. *Parasitology* **17**, 194–200.

Leigh-Sharpe, W. H. (1925b). *Lernaea* (*Lernaeocera*) *elegans* n. sp., a parasitic copepod of *Anguilla japonica*. *Parasitology* **17**, 245–251.

Lewis, A. G. (1961). "A contribution to the biology of caligoid copepods parasitic on acanthurid fishes of the Hawaiian Islands." Ph.D. Thesis, University of Hawaii.

Lewis, A. G. (1963). Life history of the caligid copepod *Lepeophtheirus dissimulatus* Wilson, 1905 (Crustacea: Caligoida). *Pacific Science* **17**, 192–242.

Lewis, A. G. (1966). The sternal furca of caligoid copepods. *Crustaceana* **10**, 7–14.

Lewis, A. G. (1969). A discussion of the maxillae of the "Caligoida" (Copepoda). *Crustaceana* **16**, 65–77.

Lewis, A. G., Dean J. and Gilfillan, E. III. (1969). Taxonomy and host associations of some parasitic copepods (Crustacea) from pelagic teleost fishes. *Pacific Science* **23**, 414–437.

Logan, V. H. and Odense, P. H. (1974). The integument of the ocean sunfish (*Mola mola* L.) with observations on the lesions from two ectoparasites, *Capsala martinieri* (Trematoda) and *Philorthagoriscus serratus* (Copepoda). *Canadian Journal of Zoology* **52**, 1039–1045.

Lopez, G. (1976). Redescription and ontogeny of *Lepeophtheirus kareii* Yamaguti, 1936 (Copepoda: Caligoida). *Crustaceana* **31**, 203–207.

Lüling, K. H. (1953). Gewebschäden durch parasitäre Copepoden besonders durch *Elytrophora brachyptera*. *Zeitschrift für Parasitenkunde* **16**, 84–92.

Manier, J.-F., Raibaut, A. Rousset, V. and Coste, F. (1977). L'appareil génital mâle et la spermiogenèse du copépode parasite *Naobranchia cygniformis* Hesse, 1863. *Annales des Sciences Naturelles, Zoologie et Biologie Animale* (12 série), **19**, 439–458.

Mann, H. (1952–1953). *Lernaeocera branchialis* (Copepoda parasitica) und seine Schadwirkung bei einigen Gadiden. *Archiv für Fischereiwissenschaft* **4**, 133–143.

Markevich, A. P. (1964). [On phylogenetic relationships of Copepoda parasitica.] *Trudy Ukrainskogo Respublikanskogo Nauchnogo Obshchestva Parazitologov*, No. 3, 3–8. (In Russian.)

Markevich, A. P. and Titar, V. M. (1978). Copepod parasites of marine fishes from the Soviet Far East. *Proceedings of the Fourth International Congress of Parasitology*, Section H, pp. 38–39.

Mirzoeva, L. M. (1972). [Life cycle and biology of *Sinergasilus lieni* Yin, 1949 (Crustacea, Copepoda parasitica).] *Parazitologiya* **6**, 252–258. (In Russian.)

Mirzoeva, L. M. (1973). [Life cycle and morphology of *Sinergasilus lieni* Yin, 1949 (Copepoda parasitica).] *Trudy Vsesoyuznogo Nauchno-issledovatelskogo Instituta Prudovogo Rybnogo Khozyaystva* **22**, 143–158. (In Russian.)

Monterosso, B. (1923). Contributo allo studio di *Peroderma cylindricum*. *Atti della Accademia delle Scienze Naturali in Catania* **13**, 1–19.

Monterosso, B. (1925). Sur la struttura e la funzione delle appendici rizoidi cefaliche di *Peroderma cylindricum* Heller. *Bollettino della Accademia delle Scienze Naturali in Catania* **54**, 3–8.

Monterosso, B. (1926). Contributo allo cognoscenza dei copepodi parassiti. Le appendici rizoidi di "*Peroderma cylindricum*" Heller. *Archivio di Biologie* **36**, 167–223.

Moser, M. and Taylor, S. (1978). Effects of the copepod *Cardiodectes medusaeus* on the lanternfish *Stenobrachius leucopsaurus* with notes on hypercastration by the hydroid *Hydrichthys* sp. *Canadian Journal of Zoology* **56**, 2372–2376.

Musselius, V. A. (1967). [Contribution to the knowledge of biology and specificity of *Sinergasilus lieni* Yin, 1949 (Crustacea, Copepoda).] *Parazitologiya* **1**, 158–160. (In Russian.)

Natarajan, P. and Nair, B. N. (1972). Observations on the nature of attack of *Lernaeenicus hemirhamphi* Kirtisinghe on *Hemirhamphus xanthopterus* (Val.). *Journal of Animal Morphology and Physiology* **20**, 56–63.

Noble, E. R., King, R. E. and Jacobs, B. L. (1963). Ecology of the gill parasites of *Gillichthys mirabilis* Cooper. *Ecology* **44**, 295–305.

Nordmann, A. von (1832). "Mikrographische Beiträge zur Naturgeschichte der wirbellosen Thiere," Heft 2. G. Reimer, Berlin.

Nunes-Ruivo, L. (1966). La genre *Alella* Leigh-Sharpe, 1925 (Copepoda, fam. Lernaeopodidae). *Proceedings of the First International Congress of Parasitology* **2**, 1081–1082.

Pannikar, N. K. and Sproston, N. G. (1941). Osmotic relations of some metazoan parasites (*Lernaeocera, Bopyrus*). *Parasitology* **33**, 214–223.

Paperna, I. and Zwerner, D. E. (1976). Studies on *Ergasilus labracis* Krøyer (Cyclopidea: Ergasilidae) parasitic on striped bass, *Morone saxatilis*, from the lower Chesapeake Bay. I. Distribution, life cycle and seasonal abundance. *Canadian Journal of Zoology* **54**, 449–462.

Pillai, N. K. (1970). A few examples showing the effect of hosts on the morphological characters of copepod parasites. *Journal of Parasitology* **56**, Section II, Part 1, 267.

Poddubnaya, A. V. (1973). [Variability and specificity of *Lernaea* parasitic on pond fishes.] *Trudy Vsesoyouznogo Nauchno-issledovalskogo Instituta Prudovogo Rybnogo Khozyaystva* **22**, 159–173. (In Russian.)

Poddubnaya, A. V. (1978). [Contribution to the knowledge of zoogeography of the crustacean genus *Lernaea* Linné, 1746.] *Trudy Vsesoyouznogo Nauchno-issledovatelskogo Instituta Prudovogo Rybnogo Khozyaystva* **27**, 111–124. (In Russian.)

Polyanski, Yu. I. (1961). Ecology of parasites of marine fishes. *In* "Parasitology of fishes" (V. A. Dogiel, G. K. Petrushevski and Yu. I. Polyanski, eds), pp, 44–83. Oliver and Boyd, Edinburgh and London.

Rawson, N. V. Jr. (1977). Population biology of parasites of striped mullet, *Mugil cephalus* L. Crustacea. *Journal of Fish Biology* **10**, 441–451.

Reddiah, K. (1970). Fecundity and its significance in parasitic copepods. *Journal of Parasitology* **56**, Section II, Part 1, 278.

Rigby, D. W. and Tunnel, N. (1971). Internal anatomy and histology of female *Pseudocharopinus dentatus* (Copepoda, Lernaeopodidae). *Transactions of the American Microscopical Society* **90**, 61–71.

Rogers, W. A. (1968). The biology and control of the anchor worm, *Lernaea cyprinacea*. *FAO Fisheries Reports* **44**, 393–398.

Rogers, W. A. (1969). *Ergasilus cyprinaceus* sp. n. (Copepoda: Cyclopoida) from cyprinid fishes of Alabama, with notes on its biology and pathology. *Journal of Parasitology* **55**, 443–446.

Rousset, V., Raibaut, A. Manier, J.-F. and Coste, F. (1978). Reproduction et sexualité des copépodes parasites de poissons. I. L'appareil reproducteur de *Chondracanthus angustatus* Heller, 1865; anatomie, histologie et spermiogenèse. *Zeitschrift für Parasitenkunde* **55**, 73–89.

Russel, F. S. (1933). On the occurrence of young stages of Caligidae on pelagic young fishes in the Plymouth area. *Journal of the Marine Biological Association of the United Kingdom* **18**, 551–553.

Sars, G. O. (1903). Copepoda Calanoida. In "An account of the Crustacea of Norway, with Short Descriptions and Figures of all the Species", volume 4, pp. I–IV, 1–171.

Schram, T. A. (1979). The life history of the eye-maggot of the sprat, *Lernaeenicus sprattae* (Sowerby) (Copepoda, Lernaeoceridae). *Sarsia* **64**, 279–316.

Schuurmans Stekhoven, J. H. Jr. (1934). Zur Sinnesphysiologie der parasitären Copepoden *Lepeophtheirus pectoralis* (O. F. Müller) und *Acanthochondria depressa* (T. Scott). *Zeitschrift für Parasitenkunde* **7**, 336–362.

Schuurmans Stekhoven, J. H. Jr. (1936). Beobachtungen zur Morphologie und Physiologie der *Lernaeocera branchialis* und *L. lusci*. *Zeitschrift für Parasiten-kunde* **8**, 659–696.

Scott, A. (1901). *Lepeophtheirus* and *Lernaea*. *Liverpool Marine Biological Committee Memoirs on Typical British Marine Plants and Animals* No. 6, 1–54.

Scott, A. (1929). The copepod parasites of Irish Sea fishes. *Proceedings and Transactions of the Liverpool Biological Society* **43**, 81–119.

Scott, T. and Scott, A. (1913). "The British parasitic Copepoda", volumes 1 and 2. Ray Society, London.

Shields, J. R. and Goode, R. P. (1978). Host rejection of *Lernaea cyprinacea* L. (Copepoda). *Crustaceana* **35**, 301–307.

Shields, J. R. and Sperber, R. G. (1974). Osmotic relations of *Lernaea cyprinacea* (L.) (Copepoda). *Crustaceana* **26**, 157–171.

Shields, J. R. and Tidd, W. M. (1968). Effect of temperature on the development of larval and transformed females of *Lernaea cyprinacea* L. (Lernaeidae). *Crustaceana* Supplement. 1, 87–95.

Shotter, R. A. (1971). The biology of *Clavella uncinata* (Müller) (Crustacea: Copepoda). *Parasitology* **63**, 419–436.

Smith, J. W. (1969). The distribution of one monogenean and two copepod parasites of whiting, *Merlangius merlangus* (L.), caught in British waters. *Nytt Magasin for Zoologi* **17**, 57–63.

Sproston, N. G. (1942). The developmental stages of *Lernaeocera branchialis*. *Journal of the Marine Biological Association of the United Kingdom* **25**, 441–466.

Srinivasachar, H. R. and Shakuntala, K. (1975). Ecophysiology of a host-parasite system: effect of infection of a parasitic copepod *Lernaea hesaragattensis* on the oxygen consumption of the fish, *Lebistes reticulatus* Peters. *Current Science (Bangalore)* **44**, 51–52.

Sundnes, G. (1970). "*Lernaeocera branchialis* (L.) on cod (*Gadus morhua* L.) in Norwegian waters." Institute of Marine Research, Bergen.

Tedla, S. and Fernando, C. H. (1970). On the biology of *Ergasilus confusus* Bere, 1931 (Copepoda), infesting yellow perch, *Perca fluviatilis* L. in the Bay of Quinte, Lake Ontario, Canada. *Crustaceana* **19**, 1–14.

Thurston, J. P. (1969). The biology of *Lernaea barnimiana* (Crustacea: Copepoda) from Lake George, Uganda. *Revue de Zoologie et de Botanique Africaines* **80**, 15–33.

Titar, V. M. (1975). [*Markevitchielinus anchoratus* gen. et sp. nov.—a genus and species of the family Chondracanthidae H. Milne Edwards, 1840.] In ["Parasites of animals and man"] (B. N. Mazurmovich, ed.). Naukova Dumka, Kiev. (In Russian.)

Vaissière, R. (1961). Morphologie et histologie comparées des yeux des crustacés copépodes. *Archives de Zoologie Expérimentale et Générale* **100**, 1–125.

Voth, D. R. (1972). Life history of the caligid copepod *Lepeophtheirus hospitalis* Fraser, 1920 (Crustacea: Caligoida). *Dissertation Abstracts International, B: Science and Engineering* **32**, 5547–5548.

Walkey, M., Lewis, D. B. and Dartnall, H. J. G. (1970). Observations on the host-parasite relations of *Thersitina gasterostei* (Crustacea: Copepoda). *Journal of Zoology (London)* **163**, 371–381.

Wierzejski, A. (1877). Ueber Schmarotzerkrebse von Cephalopoden. *Zeitschrift für Wissenschaftliche Zoologie* **29**, 562–582.

Wilkes, S. N. (1966). "The life history of *Nectobrachia indivisa* Fraser, 1920 (Copepoda: Lernaeopodoida), a parasite of the starry flounder." Ph.D. Thesis, Oregon State University.

Wilson, C. B. (1905). North American parasitic copepods belonging to the family Caligidae. Part 1. Caliginae. *Proceedings of the United States National Museum* **28**, 479–672.

Wilson, C. B. (1910). The classification of copepods. *Zoologischer Anzeiger* **35**, 609–620.

Wilson, C. B. (1911). North American copepods. Part 9. The Lernaeopodidae. Development of *Achtheres amblopilitis* Kellicott. *Proceedings of the United States National Museum* **39**, 189–226.

Wilson, C. B. (1916). Copepod parasites of freshwater fishes and their economic relation to mussel glochidia. *Bulletin of the United States Bureau of Fisheries* (1914) **34**, 333–374.

Wilson, C. B. (1922). North American parasitic copepods belonging to the family Dichelesthiidae. *Proceedings of the United States National Museum* **90**, 1–100.

Wilson, C. B. (1932). The copepods of the Woods Hole region, Massachusetts. *Bulletin of the United States National Museum* **158**, I–XIX and 1–653.

Yamaguti, S. (1963). "Parasitic Copepoda and Branchiura of fishes." Interscience Publishers, New York, London and Sydney.

Zmerzlaya, E. I. (1972). [*Ergasilus sieboldi* Nordmann, 1832, its development, biology and epizootiological significance.] *Izvestiya Gosudartvennogo Nauchno-issledovatelskogo Instituta Ozernogo i Rechnogo Rybnogo Khozyaystva* **80**, 132–177. (In Russian.)

Aspects of Acanthocephalan Reproduction

V. R. PARSHAD

Department of Zoology, Punjab Agricultural University, Ludhiana 141004, India

AND

D. W. T. CROMPTON

Molteno Institute, University of Cambridge, Downing Street, Cambridge CB2 3EE, England

I. INTRODUCTION

More than a decade ago, Schmidt (1969) estimated that about 1000 species of acanthocephalan worm had been described. All are endoparasites which are believed to attain maturity in the alimentary tract of vertebrates. Their life cycles, where known, are indirect and have been found to involve an arthropod intermediate host. Adult acanthocephalan worms may be recognized by the possession of an eversible proboscis, which generally bears hooks, a muscular proboscis sheath or receptacle, a pair of lemnisci, a typical body wall and the absence of an alimentary tract. The male and female worms possess highly characteristic reproductive organs which are suspended in the body cavity. Sources of more information about the functional morphology, life cycles, physiology and host–parasite relationships of the Acanthocephala have been compiled by Meyer (1933), Van Cleave (1953), Petrochenko (1956, 1958), Bullock (1969), Nicholas (1967, 1973) and Crompton (1970, 1975). Following Van Cleave (1948), Hyman (1951) and Bullock (1969) the Acanthocephala are treated in this review (Table 1) as a separate phylum consisting of three orders, Palaeacanthocephala, Archiacanthocephala and Eoacanthocephala (Bullock, 1969).

Reproduction is difficult to define precisely. It is more than multiplication or breeding and cannot be considered to have occurred until a population of parents has produced another, similar but later, population of parents (Cohen, 1977). This concept draws attention again to ideas about germ plasm and soma and it is perhaps helpful initially to consider acanthocephalan reproduction in the following terms. Germ cells give rise to gametes which fuse to form a zygote and this subsequently produces more germ cells which will form more gametes. Meanwhile, the zygote also gives rise to the soma or body which carries and nourishes the germ cells. The properties of the body and its responses to environmental conditions and stimuli will largely determine whether or not the gametes will actually ever have any chance of meeting with others. At present, there are many gaps in our knowledge of acanthocephalan reproduction both in terms of detail for individual species and of comparative features within the phylum. It appears, however, that acanthocephalan reproduction depends on heterosexuality followed by the active transfer of the male gamete by the male to the female. There are no obvious references in the literature to the existence of intersexes, hermaphrodites or parthenogenetic individuals. From the point of view of research, the likelihood that acanthocephalans depend entirely on one kind of reproduction may seem to be a convenient simplification, but this is offset by the fact that development and somatic activity and interaction can only occur naturally inside a living and changing environment. Consequently, attempts to investigate various features of the reproduction of the group are impeded by many technical difficulties.

Our objectives in preparing this review have been to bring together current knowledge of how one population of acanthocephalan parents develops from the preceding population and to identify problems for reinvestigation

TABLE 1

Species of Acanthocephala discussed in this review

Order ARCHIACANTHOCEPHALA
Macracanthorhynchus hirudinaceus
    (*Mac.*)[a]
Mediorhynchus centrorum
  Med. grandis (= *Heteroplus grandis*)
Moniliformis cestodiformis
  M. clarki
  M. dubius (= *M. moniliformis*)
Oligocanthorhynchus microcephala
Prosthenorchis sp.
Order PALAEACANTHOCEPHALA
Acanthocephalus jacksoni
  A. lucii
  A. parksidei
  A. ranae
Bolbosoma nipponicum
  B. turbinella
Centrorhynchus corvi
  C. elongatus
  C. kuntzi
  C. milvus
  C. spilornae
  C. turdi
Corynosoma bipapillum
  Cor. constrictum
  Cor. hamanni
  Cor. wegeneri
Echinorhynchus gadi
  E. lageniformis
  E. truttae
Fessisentis fessus
  F. necturorum
  F. vancleavei
Fillicolis anatis (*Fil.*)
Gorgorhynchus clavatus
Illiosentis furcatus
Leptorhynchoides thecatus
Polymorphus minutus
  P. trochus
Pomphorhynchus laevis (*Pom.*)
Porrorchis hylae
Prosthorhynchus formusos (*Pros.*)
Pseudoporrorchis centropi
  Pse. rotundatus
Rhadinorhynchus pristis
Serrasentis socialis
Sphaerechinorhynchus rotundo-
    capitatus

Order EOACANTHOCEPHALA
Acanthogyrus partispinus
Acanthosentis antispinus
  Aca. dattai
  Aca. oligospinus
  Aca. tilapiae
Hexaspiron nigericum
Neoechinorhynchus agilis
  N. buttnerae
  N. chrysemydis
  N. cristatus
  N. cylindratus
  N. emydis
  N. emyditoides
  N. pseudemydis
  N. rutili
  N. saginatus
  N. tylosuri
Octospinifer macilentis
Octospiniferoides chandleri
Pallisentis golvani (*Pal.*)
  P. nagpurensis
Paulisentis fractus (*Pau.*)
Tanaorhamphus sp.
Tenuisentis niloticus

[a] where necessary the generic abbreviations used in the text are shown

or new research. We do not claim to have examined all the relevant literature, but we hope that sufficient papers have been cited to introduce readers to many interesting aspects of acanthocephalan reproductive biology all of which must be considered in the context of endoparasitism.

II. Sexual Dimorphism

Differences in various external features exist between the sexes of acanthocephalan worms in their definitive hosts. Sexual dimorphism is a common but not a necessary feature of sexually reproducing organisms. As the female becomes more adapted to receive the male and then to process the products of fertilization, and the male becomes more adapted to bring the spermatozoa to the female, differences begin to occur and these may be reinforced by sexual selection (Calow, 1978). Sexual dimorphism may be frequently observed in the size and shape of the body, the distribution of body spines, the size, shape and hooks of the proboscis, the occurrence of papillae and the position of genital openings (Van Cleave, 1920; Ward and Nelson, 1967; see Yamaguti, 1963). The most obvious difference between the sexes is that of body size. In 76 out of 79 reasonably complete taxonomic descriptions, which were compiled by Yamaguti (1963), female acanthocephalans are recorded as being longer than males, although in 32 of these descriptions there is some overlap in the range of lengths of the male and female worms. *Moniliformis dubius*, *Heteroplus grandis* and *Hexaspiron nigericum* are three species in which mature females are reckoned to be about five times as long as mature males (Yamaguti, 1963). *Corynosoma hamanni* (Holloway and Nickol, 1970) and *Echinorhynchus lageniformis* (Olson and Pratt, 1971) are rather exceptional species in which the males have been observed to be bigger than the females.

Most measurements of body size are based on worms from natural infections. Variations in the lengths or body sizes of the sexes of some species are known to be associated with the age (Crompton, 1972a), reproductive state (Crompton, 1974) and population structure of the worms (Graff and Allen, 1963; Nesheim *et al.*, 1978) and to several other factors including host species and distribution (Bullock, 1962; Amin, 1975b; Buckner and Nickol, 1979), host sex (Graff and Allen, 1963), host diet (Nesheim *et al.*, 1977, 1978; Parshad *et al.*, 1980a) and host environment (Walkey, 1967). Some of these relationships have been explored experimentally with *M. dubius* in laboratory rats. Graff and Allen (1963) found that the mean fresh weights of male and female *M. dubius* were usually greater when the worms had grown in male rather than female rats and, in the absence of females, male *M. dubius* appeared to grow bigger than in their presence. Crompton (1972a) observed that male and female *M. dubius* grew at a similar rate and contained about the same amount of nitrogen (protein) until a time which probably coincided with the onset of copulation; thereafter, worms of both sexes continued to grow, although the females grew more than the males regardless of whether insemination had occurred or not (Crompton, 1974). Similarly, Crompton and Whitfield (1968) noted that the growth of male and female *Polymorphus minutus* in ducks was similar until the time when copulation first appeared to occur. Nesheim *et al.* (1977, 1978) and Parshad *et al.* (1980a) demonstrated that the body lengths and body masses of male and female *M. dubius* could be affected by the quality and quantity of carbohydrate ingested by the rat and by the numbers of worms present in the small intestine (Nesheim *et al.*, 1978). Buckner and Nickol (1975) found that the degree of sexual dimorphism in *M. clarki* and *M. moniliformis* was greater when the worms were allowed to grow in hamsters than in ground squirrels and laboratory rats respectively. At present, therefore, it is difficult to see how as variable and plastic a character as body size could contribute initially to recognition between male and female acanthocephalans during sexual congress, particularly since the little available experimental evidence suggests that the dimorphism of this character may not always become prominent until after the occurrence of copulation when the next generation of parents can be seen to be functional.

In 1920, when fewer species of Acanthocephala were known than now, Van Cleave was very cautious about assuming that somatic differences between the sexes might be involved in sexual selection. Nevertheless, it is still possible that some of the differences in body shape and form, which are seen in many species (see Yamaguti, 1963), may be connected with sex recognition. For example, the bodies of male *E. lageniformis* are cylindrical in shape whereas those of the females are seen to be bulbous (Olson and Pratt, 1971). The posterior ends of *Porrorchis hylae* (Schmidt and Kuntz, 1967) and *Fessisentis fessus* (Nickol, 1972) are dilated in females and not in males. A finger-like process is present at the posterior end of female, but not in male *Centrorhynchus kuntzi* (Schmidt and Neiland, 1966). The metasomal body of *Cor. hamanni* bears trunk spines which are distributed circumferentially on the anterior part of the trunk of both sexes (Holloway and Nickol, 1970). The posterior distribution of the trunk spines in the male *Cor. hamanni* differs from that of the female in that the genital opening of the male at the posterior end is entirely surrounded by spines whereas in the female, the spines are limited to the ventral surface in the same region. Marked differences in the form of the proboscis are known to occur between mature male and female *Filicollis anatis*, but not between very young forms in their definitive host (Van Cleave, 1920). The proboscis of a mature female *Fil. anatis* forms a globular bulb at the apex (Van Cleave, 1920) whereas that of a female *P. trochus* (Schmidt, 1965a) is swollen at the base. Scrutiny of the references listed by Yamaguti (1963) would provide many more examples of large and small external morphological differences between male and female acanthocephalans of the same species. It is tempting to speculate that small morphological differences at the posterior ends of the body, where copulation occurs (see Sections IV B and VI B), are more likely to facilitate sexual recognition than any other morphological character. Unfortunately, there does not appear to be any evidence to indicate whether the production and detection of chemical factors are involved in sex recognition. Abele and Gilchrist (1977) have implied that copulation in *M. dubius* is probably not an indiscriminate happening because when it occurs between males there is no evidence of insemination.

TABLE 2

Chemical differences between male and female acanthocephalans

Species	Content (% of wet weight)		References
	Male	Female	
LIPID			
Macracanthorhynchus hirudinaceus (Ar.)	1·7	0·9	Beames and Fisher (1964)
Moniliformis dubius (Ar.)	7·2	4·2	Beames and Fisher (1964)
Centrorhynchus corvi (Pa.)	5·1	4·3	Parshad and Guraya (1977a)
POLYSACCHARIDE			
Mac. hirudinaceus	2·3	1·4[a]	Von Brand (1940)
M. dubius (male rats)	1·6–1·7[b]	0·7–0·8	
M. dubius (female rats)	2·0	1·2	Graff and Allen (1963)

[a] About 80% of this amount is in the body wall and 12% in the reproductive tract (Von Brand, 1939).

[b] In a single-sex infection in male rats, about 4·1% of the wet weight was polysaccharide. Abbreviations: Ar., Archiacanthocephala; Pa., Palaeacanthocephala.

Morphological differences, in addition to those involving size, sometimes occur between internal non-reproductive structures of male and female acanthocephalans of the same species (see Bullock, 1962). Chemical and metabolic differences between the sexes of the same species have also been observed (Table 2). Studies *in vitro* of amino acid transport by *Macracanthorhynchus hirudinaceus* have shown that the uptake of serine and alanine by males, but not by females, is inhibited in the presence of methionine (Rothman and Fisher, 1964). Male *M. dubius* were observed to absorb more glucose per unit weight than females of the same age (Graff, 1964). These examples of chemical (Table 2) and physiological differences may be related to the metabolic needs of the worms while carrying out their roles in reproduction.

III. Sex Determination and Sex Ratio

From the available information, it appears that the sex of acanthocephalan worms, like that of many other dioecious organisms (Sinnott *et al.*, 1958), is established at fertilization by a mechanism involving sex chromosomes (Table 3). So far, two types of sex determination have been identified in which the males are heterogametic (XO or XY) and the females are homogametic (XX). At present, the mechanism involving XO appears to be the commoner of the two and is represented in each order of the phylum (Table 3).

Robinson (1965), during a study of the cytology of the testes and ovarian balls of 21-day-old male and female *M. dubius*, observed four homologous pairs of chromosomes at oogonial mitotic metaphase ($2n=8$). He identified a small pair of metacentric chromosomes which measured about 4 μm in length and three other metacentric pairs which were about 7 μm long. On examination of spermatogonial metaphase, he recognized seven chromosomes, there being three pairs and one solitary chromosome ($2n=7$). Thus, he concluded that an XO sex-determining mechanism operated in *M. dubius* with the male as the heterogametic sex. Robinson assumed that the fourth pair of chromosomes, which were seen at oogonial metaphase, were X chromosomes because of their close morphological resemblance to the unpaired chromosomes seen at mitotic metaphase in the spermatogonia. The karyotypes of *Echinorhynchus truttae* (Parenti *et al.*, 1965), *Leptorhynchoides thecatus* (Bone, 1974a) and *Neoechinorhynchus cylindratus* (Bone, 1974b) have been examined and a similar sex-determining mechanism has been proposed.

TABLE 3

Chromosomes and sex determination in some Acanthocephala

Species	Chromosome number ($2n$)	Sex chromosomes	References
PALAEACANTHOCEPHALA			
Echinorhynchus gadi (f)[a]	16	—	Walton (1959)
E. truttae (f)	M=7, F=8[b]	XO–XX	Parenti *et al.* (1965)
Pomphorhynchus laevis (f)	8	—	Walton (1959)
Acanthocephalus ranae (a)	16	—	Walton (1959)
Leptorhynchoides thecatus (f)	M=5, F=6	XO–XX	Bone (1974a)
Polymorphus minutus (b)	16	—	see Bone (1974b)
	6 or 12	—	Nicholas and Hynes (1963)
ARCHIACANTHOCEPHALA			
Macracanthorhynchus hirudinaceus (m)	M=6, F=6[c]	XY–XX	Jones and Ward (1950) Robinson (1964)
Moniliformis dubius (m)	M=7, F=8	XO–XX	Robinson (1965)
Mediorhynchus grandis (b)	6	—	Schmidt (1973)
EOACANTHOCEPHALA			
Neoechinorhynchus cylindratus (f)	M=5, F=6	XO–XX	Bone (1974b)

[a] Definitive host: a, amphibian; b, bird; f, fish; m, mammal.
[b] M, male; F, female.
[c] Chromosomes were observed to vary in length with some as long as 14 μm.

Earlier, Robinson (1964) had examined the germ cells of mature male and female *Mac. hirudinaceus* of unknown age. At oogonial mitotic metaphase, he observed three homologous pairs of chromosomes ($2n=6$) and at spermatogonial mitotic metaphase three pairs were also present ($2n=6$). One acrocentric pair and one metacentric pair were similar in appearance to pairs in the female, but the third pair in the male consisted of a metacentric chromosome with arms of equal length and a subacrocentric chromosome with arms markedly different in length. Robinson decided that the meta-centric and subacrocentric members of this third pair were the X and Y chromosomes respectively. Jones and Ward (1950) had also proposed an XY sex-determining mechanism for *Mac. hirudinaceus*, but their descriptions of the chromosomes differ from those of Robinson. The karyotypes of more species of Acanthocephala need to be determined in order to evaluate the taxonomic and phylogenetic significance of the XY and XO mechanisms. Presumably, the XO condition could have been derived from the XY (Robinson, 1965); instances of the loss of Y chromosomes from animals are not uncommon (Lewis and John, 1968).

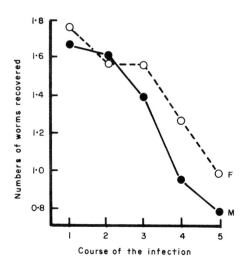

Course of the infection

FIG. 1. Evidence indicating that female (F,○) *Moniliformis dubius* (Archiacanthocephala) survive longer in rats than males (M,●). The periods represent groups of 4 or 5 weeks and the numbers of worms are represented by a transformed variable. (Redrawn from Fig. 2b, Crompton and Walters (1972), *Parasitology* **64**, 517.)

Mechanisms of sex determination which depend on one heterogametic and one homogametic parent ought in theory to give rise to a sex ratio of 1 at the time of fertilization, unless one of the two types of gamete from the heterogametic sex is in some way more active or favoured during fertilization. Observations on natural and experimental worm populations indicate that in *M. dubius* (Burlingame and Chandler, 1941; Graff and Allen, 1963; Crompton and Walters, 1972), *Mac. hirundinaceus* (Kates, 1944), *P. minutus* (Nicholas and Hynes, 1958; Crompton and Whitfield, 1968), *E. truttae* (Parenti *et al.*, 1965; Awachie, 1966), *N. saginatus* (Muzzal and Bullock, 1978) and *F. necturorum* (Nickol and Heard, 1973) a sex ratio of about 1 probably exists for the first part of the course of the infection in the definitive host. Further observations by the same authors on the same species, which represent the three orders of the phylum, also indicated that the male worms did not usually live as long as the females (Fig. 1). This difference in the life span of the sexes may have little effect on the capacity of the female worm to produce the maximum number of eggs. Crompton (1974) demonstrated, by means of experiments involving the surgical transfer of female *M. dubius* after various periods of contact with males, that a full period of egg production did not occur unless the different sexes had been together for at least 5 weeks from the beginning of the course of the infection. The sex ratio of *M. dubius*, however, was observed to remain constant during the first 5 weeks of a primary infection in laboratory rats and not to be seriously disturbed until a further 5 weeks had passed (Crompton and Walters, 1972). In apparent contrast to the generalization that the acanthocephalan sex ratio usually starts with a value of 1 before tilting in favour of females are the observations of Kennedy (1972) on *Pomphorhynchus laevis* in experimentally infected goldfish. He found that male *Pom. laevis* were unable to become established in the fish as easily as females, but once established males survived better so that the sex ratio was in favour of males after about 2 weeks.

IV. Development and Functional Morphology of the Reproductive System

The fully developed reproductive tract of a male acanthocephalan usually consists of a pair of testes and associated ducts, an assemblage of cement glands and other accessory structures, an intromittent organ and a copulatory apparatus (see Figs. 3–4). The reproductive tract of a female consists of ovarian tissue and an efferent duct system through which spermatozoa enter and eggs leave the body (see Figs. 5–7). In both sexes, the gonads are bathed in the fluid of the body cavity and are associated with the ligament strand. The early basis of our knowledge of the acanthocephalan reproductive tract is to be found in the writings of Kaiser (1893), Rauther (1930) and Meyer (1933). General information about the male and female reproductive tracts may be extracted from the taxonomic descriptions which have been compiled by Meyer (1933), Petrochenko (1956, 1958) and Yamaguti (1963); the extensive papers of Golvan, Travassos and Van Cleave are of much general relevance. The reproductive organs, and particularly those of male worms, have long been studied by taxonomists, but unfortunately from the point of view of workers concerned with the functional morphology of the reproductive tract, their whole-mounted preparations and illustrations are of limited use.

A. Development in the Intermediate Host

The development of all the organs and structures of an acanthocephalan worm takes place, in the simplest life cycle, in the body of an arthropod intermediate host (Van Cleave, 1953; Nicholas, 1967; Crompton, 1970). The rate of development varies between species and in poikilothermic hosts it is related to the ambient temperature (Table 4). There are probably optimum temperatures for the development of different species. Robinson and Jones (1971) discovered that the development of the reproductive organs and other

TABLE 4

The development of acanthocephalans in intermediate hosts

Parasite	Intermediate host	Mean[1] temperature (°C)	Genital primordium[2][3] (days)	Genital rudiment (days)	Sex[4] observed (days)	Infective cystacanth (days)	References
PALAEACANTHOCEPHALA							
Echinorhynchus lageniformis (f)[5]	*Corophium spinicorne* (amphipod)	23		13	15	30	Olson and Pratt (1971)
E. truttae (f)	*Gammarus pulex* (amphipod)	17	25	37	37	82	Awachie (1966)
Leptorhynchoides thecatus (f)	*Hyalella azteca* (amphipod)	25	15	18	18	31	DeGiusti (1949)
Polymorphus minutus (b)	*Gammarus pulex* (amphipod)	17	28	35	35	63	Hynes and Nicholas (1957); Nicholas and Hynes (1963*)
Prosthorhynchus formosus (b)	*Armadillidium vulgare* (isopod)	21	22	25	25	60	Schmidt and Olsen (1964)
ARCHIACANTHOCEPHALA							
Macracanthorhynchus hirudinaceus (m)	*Cotinus nitida* (insect)	24	20	27	27	75	Kates (1943)[6]; Meyer (1928*, 1936, 1937*, 1938a*, b*)
Mediorhynchus centrurorum (b)	*Parcoblatta pensylvanica* (insect)	27	<19	19	24	47	Nickol (1977)
M. grandis (b)	*Arphia luteola* (insect)	31	11	20	20	28	Moore (1962); Schmidt (1973*)
Moniliformis clarki (m)	*Ceuthophilus utahensis* (insect)	25	<22	22	22	60	Crook and Grundmann (1964)
M. dubius (m)	*Periplaneta americana* (insect)	27	<31	31	31	58	Yamaguti and Miyata (1942); Moore (1946); King and Robinson (1967*); Nicholas (1967*); Lackie (1972)
EOACANTHOCEPHALA							
Octospinifer macilentis (f)	*Cyclocypris serena* (ostracod)	21	<15	15	20	33	Harms (1965)
Pallisentis nagpurensis (f)	*Cyclops strenuus* (copepod)	—	<9	9	11	18	George and Nadakal (1973)
Paulisentis fractus (f)	*Tropocyclops prasinus* (copepod)	22	<7	<7	7	13	Cable and Dill (1967)
Neoechinorhynchus emydis (r)	*Cypria maculata* (ostracod)	—	<17	17	19	217	Hopp (1954)
N. rutili (f)	*Cypria turneri* (ostracod)	15	27	35	35	53	Merritt and Pratt (1964); Meyer (1931*)
N. saginatus (f)	*Cypridopsis vidua* (ostracod)	25	7	9	10	16	Uglem and Larson (1969)

[1] Mean values of temperature and days estimated by the authors from information given by the authors; there is usually considerable variation in the rate of acanthocephalan development.

[2] The terms 'primordium' and 'rudiment' are used interchangeably by some authors and separately by others.

[3] The time in days at which various developmental features have been recorded by authors.

[4] Sex is usually recognized by the presence or absence of testes.

[5] Definitive host: b, bird; f, fish; m, mammal; r, reptile.

[6] Detailed accounts of variability during the rate of larval development.

[7] The ostracod is eaten by snails (*Campeloma rufum*) in which further growth of *N. emydis* occurs.

* Describes the embryology of the species.

structures in *M. dubius* in cockroaches became abnormal when the ambient temperature was set at 33°C and above. Cable and Dill (1967) observed that the development of *Paulisentis fractus* was retarded when the copepod host *Tropocyclops prasinus* was not given sufficient food.

Development begins when the egg (Section VIII) is ingested by a susceptible intermediate host, where upon the acanthor larva emerges from the egg envelopes, penetrates the gut wall, enters the body cavity and changes through a series of acanthella stages to give rise to the stage which is involved in the infection of the definitive host. Understandably, there exists a considerable variety of terms for the different acanthella stages (King and Robinson, 1967; Lackie, 1972) and the terms cystacanth (Hynes and Nicholas, 1957), juvenile (Merritt and Pratt, 1964) and larva (Olson and Pratt, 1971) are used interchangeably to describe the stage at the end of development in the intermediate host. Metamorphosis occurs during acanthocephalan development which is illustrated, with particular reference to the reproductive tract, in Fig. 2A–G for the palaeacanthocephalan *Prosthorhynchus formosus* (Schmidt and Olsen, 1964). There is general agreement that the gonadial and most of the somatic organ primordia arise during development from the central nuclear mass (Fig. 2A, c.nu.m) located in the middle of the acanthor which is the end product of embryonic development in the body cavity of the inseminated female worm (Section VIII A). Both the male and female reproductive tracts develop in close association with the ligament (Fig. 2G). A general description of acanthocephalan development has been given by Meyer (1933) and specific details may be obtained through the references cited in Table 4. The following account is concerned with the reproductive tract. The term primordium is used to describe nuclei or cells that are morphologically undifferentiated, but of known development fate. The term rudiment is used for structures that are displaying some degree of differentiation.

1. *Male reproductive tract*

According to Schmidt and Olsen (1964), the development of *Pros. formosus* in its isopod intermediate hosts takes about 60 days at a temperature of 20–23°C. After 22 days, the central nuclear mass has increased in size and the nuclei have become grouped into primordia (Fig. 2B). The primordium in the middle of the acanthella gives rise to the rudiments of the body wall musculature, the ligament and the gonads; the posterior primordium, which contains larger nuclei than the middle primordium, gives rise to the rudiments of the copulatory structures (Fig. 2B). About 3 days later (Fig. 2C), the rudimentary ligament and rudimentary testes are visible. Further differentiation is apparent in the posterior primordium where the nuclei have become organized into distinct groups. In 27-day-old male acanthellae, the testes are very prominent and the rudimentary cement glands, ducts and copulatory bursa can be distinguished (Fig. 2D). The gradual differentiation of these structures continues (Figs 2E and 2F) until the cystacanth stage is formed. Other published accounts of acanthocephalan development (Table 4) differ in points of detail and timing from this description for *Pros. formosus*. The association of the developing ligament with the development of the male gonads (and female—see below) is intriguing and in need of further investigation.

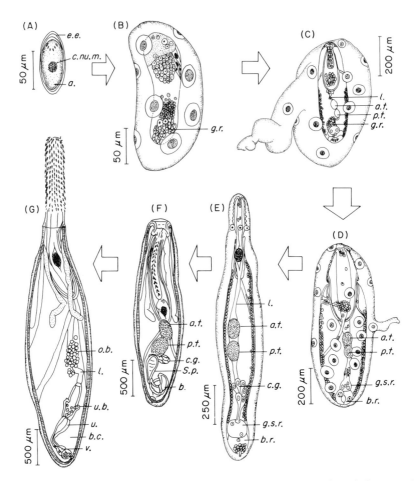

FIG. 2. Development of the reproductive system of the palaeacanthocephalan *Prostho-rhynchus formosus* in its isopod intermediate host (20–23°C). (Redrawn from Figs. 1, 3, 4, 5, 9, 10 and 11 of Schmidt and Olsen (1964), *J. Parasit.* **50,** 721.) (A) Infective egg consisting of the egg envelopes and the acanthor larva. (B) 22-day acanthella. (C) 25-day male acanthella. (D) 27-day male acanthella. (E) 30-day male acanthella. (F) Male cystacanth. (G) Female cystacanth. For details of abbreviations, see p. 85.

2. Female reproductive tract

In *Leptorhynchoides thecatus* (DeGiusti, 1949), and probably in most species, the male and female reproductive system develop along comparatively similar lines. The testes can usually be identified earlier than the ovarian

Explanation of Lettering Used in the Figures

a.	acanthor larva	*l.*	ligament
ax.	axon	*le.*	lemniscus
a.b.a.m.	anterior bursal accessory muscle	*lu.*	lumen
		l.b.n.	lateral bursal nerve
a.o.	atretic oocyte	*l.m.c.*	lateral diverticulum of muscle cap
a.t.	anterior testis		
b.	bursa (copulatory)	*l.s.*	ligament sac
b.c.	body cavity	*mi.*	mitochondrion
b.d.m.	bursal depressor muscle	*mv.*	microvillus
b.r.	bursal rudiment	*m.c.*	muscle cap
b.re.	bursal receptor	*m.o.*	mature oocyte
b.s.	bursal sac	*nu.*	nucleus
b.w.	body wall	*n.f.*	nuclear fragment
c.	cement (secretion)	*o.b.*	ovarian ball
ce.	centriole	*o.b.n.*	outer bursal nerve
ch.	chromatin	*o.s.*	oogonial syncytium
c.c.	copulation cap	*p.*	penis
c.d.	cement duct(s)	*p.g.*	pyriform gland
c.g.	cement gland(s)	*p.m.*	protrusor muscle
c.nu.m.	central nuclear mass	*p.n.*	penis nerve
c.p.re.	circumpenial receptor	*p.t.*	posterior testis
c.r.	cement reservoir	*r.*	receptor
c.s.	cytophoral stalk	*s.*	spine
d.c.	dorsal commissure	*sa.*	spermatozoon
d.i.	dense inclusion	*s.c.*	secretory cell
d.l.s.	dorsal ligament sac	*s.c.g.*	secretory lobe of cement gland
d.o.	developing oocyte		
d.r.m.c.	digitform ray of muscle cap	*S.d.*	Saefftigen's duct
e.	egg(s)	*S.p.*	Saefftigen's pouch
e.e.	egg envelope(s)	*s.s.*	supporting syncytium
fl.	flagellum	*s.s.nu.*	supporting syncytium nucleus
f.c.	flagellar cleft		
gdsb.	Grundsubstanz (see Meyer, 1933.=*s.s.*)	*t.*	testis
		t.e.	testis envelope
G.b.	Golgi body	*u.*	uterus
g.g.	genital ganglion	*u.b.*	uterine bell
g.nu.	giant nucleus	*v.*	vagina
g.p.	genital pore	*v.c.*	ventral commissure
g.r.	genital rudiment	*v.d.*	vas deferens
g.s.	genital sheath	*v.e.*	vas efferens
g.s.r.	genital sheath rudiment	*v.l.s.*	ventral ligament sac
g.sp.	genital sphincter	*v.s.*	vesicula seminalis
i.b.n.	inner bursal nerve	*z.*	zygote

tissue; in *Pros. formosus* young acanthella can be recognized as female because the gonads are not obvious rather than because of their prominence (see Schmidt and Olsen, 1964). By the time the acanthellae are 30 days old, the genital primodia of the female are visible and ovarian balls are clearly present in the cystacanth (Fig. 2G).

The ovarian balls (Section IV B2b) are one of the most interesting and unusual features of the Acanthocephala. In female worms in normal definitive hosts, the ovarian tissue is organized into ovarian balls or free-floating ovaries which are in the fluid of the body cavity either freely or loosely constrained in the ligament sacs (Bullock, 1969; Crompton and Whitfield, 1974). The ligament sacs may persist throughout the life of the worm or may rupture releasing their contents into the body cavity (Bullock, 1969).

Van Cleave (1953) considered that the ovary became fragmented in some unexplained manner to form the free-floating ovarian balls. This view is shared by most workers, but an investigation of *M. dubius* by Atkinson and Byram (1976) has drawn attention to the question of when fragmentation begins. Atkinson and Byram wrote that in *M. dubius*, "Seven to nine days after the infection of the definitive host (rat) by cystacanths, the genital primordium of the female acanthocephalan is transformed from a fragmented mass of cells into discrete ovarian balls." Earlier, Yamaguti and Miyata (1942) wrote of female *M. dubius* nearing the end of their development in the intermediate host, "Der im kontrahierten ligament immer mehr wachsende Keimstock zerfallt in Keimzellenballen verschiedener Grosse, die, direct aufeinander folgend und gegeneinander abgeplattet, eine bis zur Uterusgloke reichende Zickzacklinie bilden. Die ballen bestehen aus dicht gedrangten klein Keimzellen mit verhaltnismassig grossen Kernkorperchen; die grosseren von ihnen, die bis 70 μ lang und 42 μ breit sein konnen, zeigen eine Neigung zum Zerfall im mehrere Teilstucke." Since then, Moore (1946) studied the development of *M. dubius* in its intermediate host and wrote, "No compact ovary is present in the female, but small masses of compact cells, which are immature egg balls, may be seen free in the body cavity or in the genital ligament." Asaolu (1976) has confirmed this observation. Crompton *et al.* (1976) were able to count eight spherical ovarian balls on average in 7-day-old *M. dubius* from rats; these observations are difficult to reconcile with the view of Atkinson and Byram (1976).

Schmidt and Olsen (1964) considered that the female cystacanth of *Pros. formosus* (Fig. 2G) was precocious because the ovary had developed and broken up to form ovarian balls while the females were still in the intermediate host. These events, which have also been observed by some workers during the development of *M. dubius*, may be quite common. The occurrence of ovarian fragmentation or the presence of ovarian balls have been described from the body cavities of female cystacanths of *Fessisentis necturorum* (Nickol and Heard, 1973), from female juvenile *Neoechinorhynchus rutili* in their intermediate hosts (Merritt and Pratt, 1964), from female cystacanths of *Mediorhynchus grandis* (Moore, 1962) and from developing *E. truttae* in amphipods (Awachie, 1966). On the other hand, ovarian development seems to be less advanced by the time the cystacanth or juvenile is reached in *L. thecatus* (DeGiusti, 1949), *P. minutus* (Hynes and Nicholas, 1957), *N. cylindratus* (Uglem and Larson, 1969) and *Paulisentis fractus* (Cable and Dill, 1967). Crook and Grundmann (1964) concluded that female *M. clarki* show no indication of an ovarian primordium at any stage of development in the camel cricket, *Ceuthophilus utahensis*. This varied and sometimes conflicting selection of observations on ovarian tissue in immature worms serves to emphasize the need for more comparative studies of acanthocephalan development.

B.   FUNCTIONAL MORPHOLOGY IN THE DEFINITIVE HOST

Once acanthocephalan worms have become established in their definitive hosts, further differentiation and maturation occurs in the gonads while the growth and the start of normal functioning of the various organs and structures are the commonly observed events in the somatic part of the tract.

1. *Male reproductive system*

The general disposition of the male reproductive tract in various species is shown in Fig. 3A–E. In some, the tract extends for most of the length of the body (Fig. 3A), whereas in others it appears to be concentrated near the posterior end (Fig. 3E). The reproductive organs are usually arranged with the testes being located anterior to the cement glands which are in turn anterior to the copulatory bursa (Fig. 3A–E).

(a) *Testes and sperm ducts.* Most species usually possess two testes which are often arranged in tandem (Figs 3D and 3E) although some degree of overlap (Figs 3A and 3C) is not uncommon. The testes may be located in the anterior (Fig. 3A), middle (Fig. 3B) or posterior (Fig. 3E) parts of the body and they are frequently described as being oval, elliptical or spherical in shape (see Yamaguti, 1963). The testes of *Neoechinorhynchus tylosuri* (Yamaguti, 1939) are tubular in shape. The sizes of the testes may vary within a species (Bullock, 1962; Amin, 1975a, b) as well as between species (Fig. 3). Intra-specific variation is to be expected as the male worms grow.

Monorchic specimens have now been described from several specimens including *Centrorhynchus elongatus* (Kobayashi, 1959), *Acanthocephalus jacksoni* (Bullock, 1962), in which 14 males out of a sample of 208 had one testis only, *Cor. bipapillum* (Schmidt, 1965b), *N. oreini* (Fotedar, 1968), *M. dubius* (Crompton, 1972b), *A. parksidei* (Amin, 1975a, b) and *Fessisentis vancleavei* (Buckner and Nickol, 1978). Male *F. vancleavei* were described as usually having one filiform testis (Fig. 3B) rather than two. Crompton (1972b), by means of surgical transplantation, was able to demonstrate that one monorchic *M. dubius* was capable of copulation and successful insemination. In view of the observation that the single testis from a monorchid male is generally larger than either of the testes from a diorchic male of the same species, it is usually assumed that monorchidism arises from the fusion of their rudiments during development (see Bullock, 1962). Fotedar (1968) observed in some male *N. oreini* one large testis which was slightly demarcated in the middle and he inferred that the testes had united in some way.

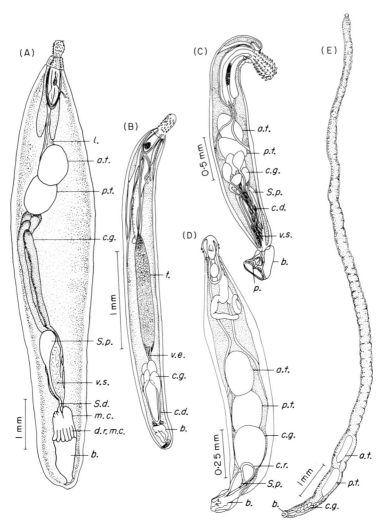

FIG. 3. Diagrammatic representations of the male acanthocephalan reproductive tract. (A) *Centrorhynchus turdi* (Pa.). (Redrawn from Fig. 13 of Yamaguti (1939), *Jap. J. Zool.* **8,** 318.) (B) *Fessisentis vancleavei* (Pa.). (Redrawn from Fig. 6 of Buckner and Nickol (1978), *J. Parasit.* **64,** 635.) (C) *Echinorhynchus lageniformis* (Pa.). (Redrawn from Fig. 12 of Olson and Pratt (1971), *J. Parasit.* **57,** 143.) (D) *Octospiniferoides chandleri* (Eo.). (Redrawn from Fig. 1 of Bullock (1969), *In* "Problems in Systematics of Parasites", University Park Press, Baltimore.) (E) Sketch of young *Moniliformis* sp. (Ar.) to show location of reproductive tract relative to rest of body. Ar., Archiacanthocephala; Pa., Palaeacanthocephala; Eo., Eoacanthocephala. For details of abbreviations, see p. 85.

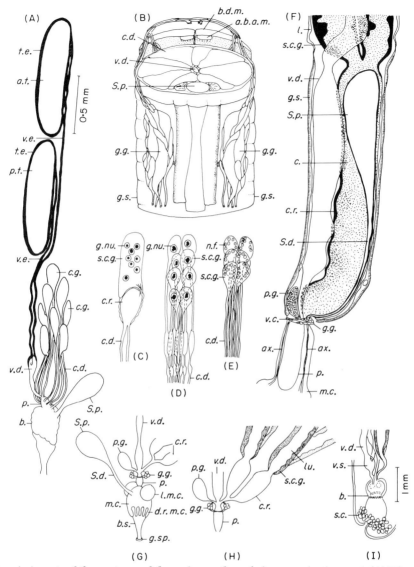

Fig. 4. Aspects of the anatomy of the male acanthocephalan reproductive tract. (A) Diagrammatic representation of the reproductive organs of *Moniliformis dubius* (Ar.). (Redrawn from Fig. 26 of Asaolu (1977), PhD. Dissertation, University of Cambridge.) (B) An illustration of the ejaculatory duct of *M. moniliformis* (Ar.) showing the relative position of the paired ganglia. (Redrawn from Fig. 2 of Dunagan and Miller (1978b), *J. Parasit.* **64,** 431.) (C) Eoacanthocephalan syncytial cement gland. (D) Archiacanthocephalan cement glands. (E) Palaeacanthocephalan cement glands. (Redrawn from Figs. A, B and C of Van Cleave (1949a), *J. Morph.* **84,** 427.) (F) Reproductive organs of *Polymorphus minutus* (Pa.) enclosed in the genital sheath. (Redrawn from Fig. 7 of Whitfield (1969), PhD. Dissertation, University of Cambridge). (G) Diagrammatic interpretation of the posterior part of the reproductive tract of *P. minutus* (Pa.). (Redrawn from Fig. 5 of Whitfield (1969), PhD. Dissertation, University of Cambridge.) (H) An interpretation of the connexions between the various parts of the cement gland apparatus of *P. minutus* (Pa.). (I) Terminal part of the reproductive tract of *Fessisentis fessus* (Pa.). (Redrawn from Fig. 2D of Dunagan and Miller (1973), *Proc. helm. Soc. Wash.* **40,** 209.) Ar., Archiacanthocephala; Pa., Palaeacanthocephala. For details of abbreviations, see p. 85.

The testes of adult *M. dubius* are enclosed in envelopes of connective tissue (Fig. 4A) which attaches them to the ligament (Asaolu, 1977, 1980). A similar arrangement probably exists in other species of Acanthocephala. A sperm duct, or vas efferens, leads from each testis and fuses posteriorly to form the vas deferens (Figs 4A, 4G–I). Some observers have used the terms seminal vesicle and vesicula seminalis for part of the sperm duct system (Figs 3A, 3C and 4I), but Yamaguti (1963) has pointed out that the seminal vesicle is often absent. There are usually two vasa efferentia in monorchic males (see Bullock, 1962). The vas deferens leads into the so-called cirrus or penis (Figs 3C, 4A, 4F–I) which is assumed to function as an intromittent organ during copulation. The common use of terms like penis, vesicula seminalis, prostatic mass (Verma and Datta, 1929), each of which was no doubt coined by mammalian anatomists, may not always be wise and may lead to some deception about our state of knowledge of the function of some of these structures. Information about testicular histology is discussed elsewhere (Section V A).

(b) *Cement and other glands.* The cement glands or glandular apparatus are structures of much taxonomic interest (Meyer, 1933; Van Cleave, 1949a; Bullock, 1969; Yamaguti, 1963). The cement glands, which may be variable in shape (Figs 3A and 3B), usually conform to one of three basic arrangements (Van Cleave, 1949a). These are the eoacanthocephalan type (Fig. 4C) in which a syncytical secretory lobe and a reservoir are present, the archiacanthocephalan type (Fig. 4D) in which eight uninucleate discrete glands are usually present, and the palaeacanthocephalan type (Fig. 4E) which may vary in number from about two to eight and in which nuclear fragments are found. These typical descriptions do not cover all known species. For example, there is structural continuity between the four cement glands of *P. minutus* (Figs 4G and 4H; Whitfield, 1969) and the six glands of *A. parksidei* (Amin, 1975a). The number of cement glands, or secretory lobes in *A. parksidei*, was found to vary from 0 (in two males) to 12 (in one male) out of a population of 1801 males. Five (in 66 males) was the commonest unusual number observed (Amin, 1975b). Van Cleave (1949b) also described a male *E. gadi* with 12 cement glands.

The secretion of the cement glands is carried posteriorly along a series of ducts (Figs 4A–4F) which either discharge into the penis, as appears to be the case in *M. dubius*, or empty into some kind of collecting chamber, as has been observed for *P. minutus* (Figs 4F–4H). Whitfield (1969) found that the secretory lobes of the cement glands of *P. minutus* consisted of a cytoplasmic wall, which was about 20 μm thick, surrounding a lumen. The cytoplasm contained membrane-bounded globules, measuring at least 1 μm in diameter, together with mitochondria, rough endoplasmic reticulum and glycogen. Asaolu (1977) made similar observations on the cement glands of *M. dubius*. Thus, these acanthocephalan cement glands are of the exocrine rather than holocrine type (Meyer, 1933). Little information is available about the nature of the cement gland secretion. The "cement" is stained an intensive blue–black colour with Heidenhain's iron haematoxylin and is likely to contain protein since it appears to originate from cytoplasm which is rich in rough endo-plasmic reticulum. Haley and Bullock (1952) implied that the cement glands of *N. emydis* and *E. gadi* secreted a proteinaceous material. The cytoplasm of *Centrorhynchus corvi* contains material which shows, like protein, affinity for orange G and mecuric bromophenol blue (V. R. Parshad, unpublished observations). In a comparative study of the distribution of alkaline phos-phatase activity in 23 species of Acanthocephala, Bullock (1958) obtained negative results for cement glands except from those of *Polymorphus* sp. and *Corynosoma wegeneri*.

The secretion or secretions of the cement glands are assumed to contribute to the copulation caps which are often observed around the posterior end of females (see Fig. 18A; Van Cleave 1949a) and sometimes on the males (Abele and Gilchrist, 1977). The caps are quite hard and the material from which they are formed has presumably become hardened since secretion. Hardening could happen on exposure to the environment or on the addition of some other agent. Whitfield (1969) described and named the pyriform gland in *P. minutus* (Figs 4G and 4H) and Asaolu (1977) observed a pair of glands in association with the cement glands of *M. dubius*. Dunagan and Miller (1973) found two types of gland-like cell which were located at the margin of the bursa of *F. fessus* (see Figs 4I and 18B). These cells were of two types; one is reniform with homogeneous contents whereas the other is spatulate with contents that appear to be granular. Dunagan and Miller suggested that these glandular cells are located in a convenient position for the release of some catalyst involved in cap formation.

(c) *Copulatory bursa*. The role of the male in copulation is achieved by means of the bursa (Figs 3A–3E) which must become fully extruded (Fig. 3C) before the female can be grasped and the spermatozoa transferred. The bursa consists of an eversible extension of the body wall; its eversion (Fig. 3C) and retraction (Fig. 3B) depending on complex musculature (Figs 3A and 4G) and perhaps on a hydraulic mechanism involving Saefftigen's pouch (Figs 3A–3D; Fig. 4F) (see Yamaguti, 1963). The pouch, reproductive ducts and the insertion points of various muscles connected with the male reproductive system are enclosed in the genital sheath (Figs 4B and 4F). A careful study of the copulatory bursa of *M. moniliformis* and a review of the observations of earlier authors by Dunagan and Miller (1978a) has provided a good under-standing of the structure and functioning of the bursa. The bursa consists of a muscular cap, a sac and a lining. The cap consists of radial muscles which are enclosed inside a loose arrangement of circular muscles. Presumably, these muscles function once the bursa is everted and is grasping the female (see Figs 18D–18G). Bursa eversion in *M. moniliformis* depends to some extent on the contraction of a relatively large sheet of muscle known as the bursal depressor muscle and on the two longitudinal bursal protrusor muscles which originate on the ventral body wall at the posterior extremity of the worm adjacent to the genital pore. The protrusor muscles are attached at their other end to the genital sheath near where it unites with the ligament. Other muscles, whose function is less well understood, include anterior bursal accessory muscles which insert on to the bursal depressor muscle. If most of this muscular tissue is involved in bursal eversion it is to be expected that some antagonistic muscles will be responsible for bursal retraction.

The exact function of Saefftigen's pouch (Fig. 4F) is still unclear. It is a fluid-filled muscular, and therefore contractile, sac which appears to make contact through Saefftigen's duct with the bursal wall (Figs 4B, 4F and 4G). Dunagan and Miller (1978a) have concluded that movement of the pouch fluid may not be important in bursal eversion. After eversion, however, an increase in the turgidity of the bursal wall resulting from an influx of fluid might help to maintain the grip of the male on the posterior end of the female.

Van Cleave (1949a) pointed out that even when the bursa is fully extruded (Fig. 3C), the penis lies internally within the dome of the bursa making insemination of the female dependent on the meticulous adjustment of the bursa around the genital extremity of the female. Furthermore, the genital pores of females of the same species may not always be in the same position (Fig. 6). More details of bursal structure and functioning and the copulatory act are given later (Section VI), but it is important to emphasize that, not surprisingly, male acanthocephalans possess a relatively complex genital nervous and sensory system (see Figs 4B, 4F–4H) in association with the copulatory bursa. Movements of the copulatory apparatus are assumed to be co-ordinated by the activity of the paired genital ganglia in *A. ranae* (Kaiser, 1893), *Macracanthorhynchus hirudinaceous* (Kaiser, 1893; Dunagan and Miller, 1979; Fig. 18E), *Oligocanthorhynchus microcephala* (Kilian, 1932), *Bolbosoma turbinella* (Harada, 1931), *P. minutus* (Whitfield, 1969) and *M. moniliformis* (Dunagan and Miller, 1977, 1978b).

2. *Female reproductive system*

Anatomical features of the female reproductive tract and organs are illustrated in Figs 5, 6 and 7. In a sexually mature inseminated female, nearly all the body cavity is occupied by the reproductive tract, the ovarian balls and eggs in various stages of development.

(a) *Ovarian tissue*. There is general agreement that the functional ovarian tissue consists of ovarian balls (Figs 5A and 8A) which float freely in and amongst the developing eggs in the fluid of the body cavity (Meyer, 1933; Van Cleave, 1953; Yamaguti, 1963; Bullock, 1969). The eggs and ovarian balls of archiacanthocephalans remain within the persistent ligament sacs (Fig. 7A), whereas in the palaeacanthocephalans (Fig. 7B) the single ligament sac disintegrates and the eggs and ovarian balls become dispersed throughout the body cavity. In eoacanthocephalans of the genera *Neoechinorhynchus*, *Octospiniferoides* (Fig. 5A) and *Tanaorhamphus* (Bullock, 1969), there are both dorsal and ventral ligament sacs as shown in Fig. 7A, but these do not persist and the eggs and ovarian balls become free in the body cavity (Bullock, 1969). "Ovarian ball" is now a well-established term presumably because it is a translation of "Ovarialballen" (Meyer, 1933), although the terms "ovarial ball", "Eiballen", "ovarian fragments", "egg balls" and "egg masses" are also used. In fact, all the available evidence shows that the mature ovarian

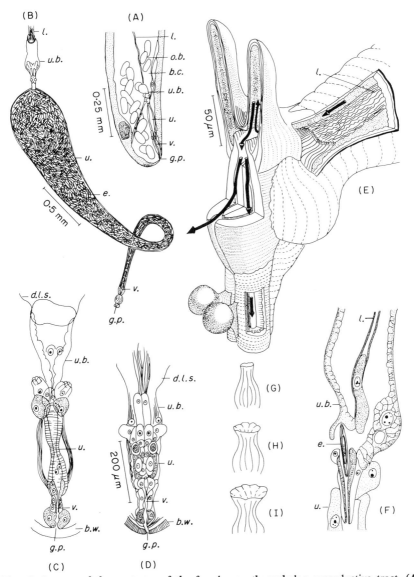

FIG. 5. Aspects of the anatomy of the female acanthocephalan reproductive tract. (A) Trunk of *Octospiniferoides chandleri* (Eo.) showing main features of the reproductive system (Redrawn from Fig. 3 of Bullock (1969), *In* "Problems in Systematics of Parasites", University Park Press, Baltimore.) (B) Efferent duct system of *Polymorphus minutus* (Pa.). (Redrawn from Fig. 1 of Whitfield (1968), *Parasitology* **58**, 671.) (C) Uterine bell of *Neo-echinorhynchus rutili* (Eo.). (Redrawn from Fig. 335b of Meyer (1933), *Acanthocephala, In* "Bronns Klassen und Ordnungen des Tierreichs".). (D) Uterine bell of *Moniliformis dubius* (Ar.). (Redrawn from Fig. 12 of Yamaguti and Miyata (1942), *Uber die Entwicklungs-geschichte von Moniliformis dubius (Acanthocephala).*) (E) Stereogram of mature uterine bell of *P. minutus* (Pa.), cut away to reveal complex internal luminal system. Possible routes for egg translocation shown by heavy arrows. (Redrawn from Fig. 9 of Whitfield (1968), *Parasitology* **58**, 671.) (F) Sagittal section through the uterine bell of *Echinorhynchus angustatus* (=*Acanthocephalus lucii*) (Pa.). (Redrawn from Pl. 7 and Fig. 15 of Kaiser (1893), *Biblthca zool.* II, Heft 7.) (G)–(I) Sketches showing different shapes adopted by the uterine bell of *Fessisentis fessus* (Pa.). (Redrawn from Fig. 2E of Dunagan and Miller (1973), *Proc. helm. Soc. Wash.* **40**, 209.) Ar., Archiacanthocephala; Eo., Eoacantho-cephala; Pa., Palaeacanthocephala. For details of other abbreviations, see p. 85.

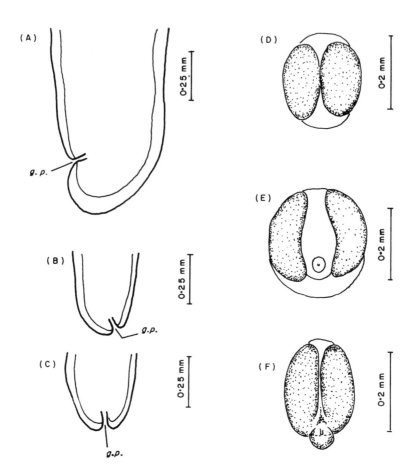

FIG. 6. Features of the posterior ends of female acanthocephalan worms. (A)–(C) Variations in the position of the genital pore of *Fessisentis fessus* (Pa.). (Redrawn from Figs. 7–9 of Nickol (1972), *J. Parasit.* **58**, 282.) (D)–(F) The posterior ends of *Neoechinorhynchus* (Eo.) *pseudemydis, N. emydis* and *N. chrysemydis* respectively. (Redrawn from Figs. 2, 9 and 15 of Fisher (1960), *J. Parasit.* **46**, 257.) Eo., Eoacanthocephala; Pa., Palaeacanthocephala. For details of other abbreviations, see p. 85.

balls are ovaries (Section V B) which should be called by that name. It is a pity that another earlier German term, "lose Ovarien" (=free ovaries) (see Meyer, 1933) was not widely adopted by English-speaking authors. Yamaguti (1939) regularly mentioned the "floating ovaries" in a paper dealing with several species of Acanthocephala.

The ovarian balls in a mature acanthocephalan are often oval in profile, because they frequently resemble prolate spheroids in shape (Fig. 8). They change from a roughly spherical to an elliptical shape in *M. dubius* during the

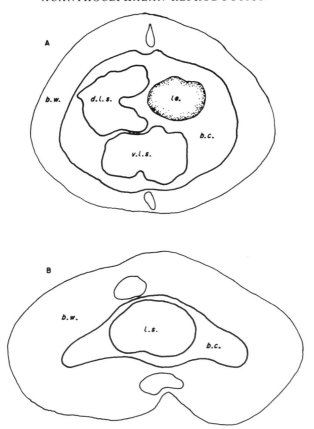

FIG. 7. Sketches of transverse sections through the female acanthocephalan body to show the arrangement of the ligament sacs. (A) Archiacanthocephalan worm and young eoacanthocephalan worm. (B) Young palaeacanthocephalan worm.

early part of the course of the infection in the rat (Crompton *et al.*, 1976). The mature ovarian balls of *P. minutus* (Fig. 8E) appear to be more spherical than those of *M. dubius* (Nicholas and Hynes, 1963; Crompton and Whitfield, 1974). Most information about the numbers and sizes of the ovarian balls in a given species is available for *M. dubius* during experimental infections in rats (Crompton *et al.*, 1976). Estimates of the average numbers of ovarian balls in both unfertilized and fertilized (=inseminated) worms are shown in Fig. 9. It appears that the maximum number of ovarian balls is reached earlier in unfertilized than fertilized worms; presumably this difference has some connexion with egg production. Eventually, however, each female worm acquires on average about 6300 ovarian balls and it was suggested by Crompton *et al.* (1976) that the increase in number may be the result of some form of division of some or all members of the ovarian ball population (Fig. 8G). Parshad *et al.* (1980b), during a further study of ovarian balls from *M. dubius*,

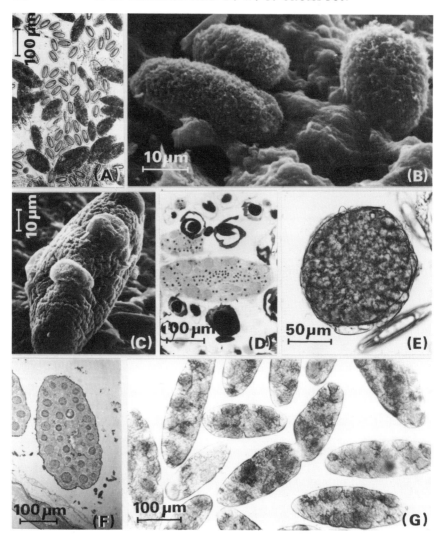

FIG. 8. Features of ovarian balls. (A) Photomicrograph of formalin-fixed ovarian balls, developing eggs and mature eggs (shelled acanthors) from the body of a fertilized female *Moniliformis dubius* (Ar.). (B) Scanning electron micrograph of ovarian balls from an immature female *M. dubius*. (C) Scanning electron micrograph of an ovarian ball from a mature female *M. dubius*. (After Pl. 1A and C of Parshad *et al.* (1980b) *Parasitology*, **81**, 423.) (D) Photomicrograph of a longitudinal section of an ovarian ball in the body cavity of a mature female *M. dubius:* Heidenhain's iron haematoxylin. (After Pl. 1B of Crompton and Whitfield (1974), *Parasitology* **69**, 429.) (E) Photomicrograph of an ovarian ball from a mature female *Polymorphus minutus* (Pa.). (F) Photomicrograph of an ovarian ball from a mature female *Centrorhynchus corvi* (Pa.). (After Pl. 1A of Parshad and Guraya (1977b), *Parasitology* **74**, 243.) (G) Photomicrograph of an ovarian ball from a mature, unfertilized female *M. dubius*. This ovarian ball might have been dividing when fixative was added. (After Pl. 1B of Crompton *et al.* (1976), *Parasitology* **73**, 65.) Ar., Archiacanthocephala; Pa., Palaeacanthocephala.

observed ovarian balls like that shown in Fig. 8G from both unfertilized and fertilized worms varying in age from 1–14 weeks. These observations also suggested that the ovarian balls do not always divide evenly, that their shape has no obvious effect on division and that the distribution of cell types within the ovarian balls (see Section V B) has no apparent effect on the location of the plane of division (Parshad *et al.*, 1980b).

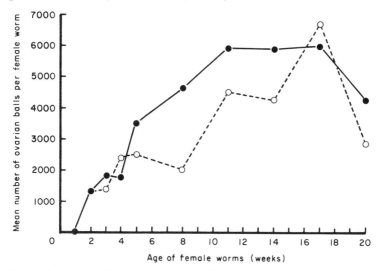

FIG. 9. Graph showing the average numbers of ovarian balls estimated to be present in unfertilized (●–●) and fertilized (○–○) female *Moniliformis dubius* (Archiacanthocephala) of varying ages during the course of the infection in male laboratory rats. (After Fig. 1 of Crompton *et al.* (1976), *Parasitology* **73**, 65.)

The ovarian balls of *M. dubius* were found to increase in size during the course of the infection, despite their increase in numbers (Fig. 9) and regardless of whether they were contained in unfertilized or fertilized female worms (Crompton *et al.*, 1976). The ovarian balls from the unfertilized worms were observed to be significantly larger on average than those from the fertilized worms. Very little information is available about the numbers and sizes of ovarian balls from other species of Acanthocephala and so it is not possible to confirm the widely held impression that the number of ovarian balls increases in all species. Recently, Parshad and Guraya (1977b) examined the complex multi-unit ovarian balls of *C. corvi* (Fig. 8F) from house crows (*Corvus splendens*) and suggested that a marked increase in the number of ovarian balls of *C. corvi* might not occur. More details of the structure and organization of the ovarian balls are considered below in the discussion of gametogenesis (Section V B) which is the primary function of the ovarian balls.

(b) *Efferent duct system.* The anatomy of the efferent duct system of *P. minutus*, which for the present may serve as the representative of a large number of acanthocephalan species, is illustrated in Fig. 5B. The inner opening of the efferent duct is to be found at the uterine bell and the outer opening forms the genital pore (Figs 5A–5D), the position of which may be variable (Fig. 6). The uterine bell is anchored to the proboscis receptaculum by the ligament, but, as was observed with males, the size of the efferent duct system relative to the size of the body of the worms is very variable between species. The uterus (Figs 5A–5D), which in mature fertilized females is usually observed to contain eggs, is located between the uterine bell and the genital pore. In addition to information about the efferent duct system in the taxonomic literature, detailed descriptions, in addition to those summarized for various species by Meyer (1933), have been given for the palaeacantho-cephalans *P. minutus* (Whitfield, 1968, 1970), *F. fessus* (Dunagan and Miller 1973) and *E. gadi* (Khatkevich, 1975a), for the eoacanthocephalan *N. rutili* by Khatkevich (1975b) and for the archiacanthocephalan *M. dubius* by Asaolu (1977).

The uterine bell (Figs 5A–5F) is a complex organ whose function has been a matter of some controversy. Usually, only mature eggs (shelled acanthors) are found in the faeces of infected definitive hosts and yet eggs at all stages of development are present in the body cavities of female worms living in the same host (Fig. 8A). Thus, some form of egg sorting is assumed to occur and the uterine bell, which is located at the inner end of the efferent duct, is considered to be involved. This suggestion for the function of the uterine bell gained support from Kaiser (1893) who studied *Acanthocephalus lucii* and depicted an egg within the uterine bell (Fig. 5F). Meyer (1933) and Yamaguti (1963) have argued that the circumstantial evidence in favour of the egg-sorting function of the uterine bell is equivocal because immature and mature eggs may be observed within the uterus. For example, Yamaguti (1939) described immature eggs inside the uterus of *Bolbosoma nipponicum* and Asaolu (1977) even saw on one occasion small ovarian balls inside the uterus of *M. dubius*. Nicholas and Grigg (1965), on the other hand, observed that almost all the eggs found in the intestinal contents of rats infected with *M. dubius* were fully developed.

Whitfield (1968), after making a thorough study of the histology and ana-tomy of the uterine bell of *P. minutus*, concluded that its morphology could best be explained in terms of egg sorting (Fig. 5E). He proceeded to test this view by estimating the population structure of the eggs, from the body cavity and the uterus, in terms of stage of development (Table 5). Hardly any eggs were observed in the uterus until the worms were beginning to release eggs when about 3 weeks old (Crompton and Whitfield, 1968) and all those that were observed in the uterus once egg release had begun appeared to be fully developed (Table 5). Whitfield also developed an *in vitro* preparation of the efferent duct of *P. minutus* which enabled him to observe the motility of the uterine bell; earlier work had shown that the walls of the uterine bell (Fig. 5E) and uterus (Fig. 5B) were muscular (Whitfield, 1968). After fortuitously observing the movement of eggs in the uterus and uterine bell *in vitro*, Whitfield (1970) concluded that the mechanism of egg sorting by the uterine bell of *P. minutus* would most probably depend on the bell distinguishing between eggs of different sizes with mature ones being at least 100 μm in length. Another aspect of Whitfield's study which supported egg sorting as a normal function of the uterine bell was his contention that immature eggs might slip into the uterus as the result of abnormal bell activity following the disturbance of the worms on the death of their hosts (Whitfield, 1970). The efferent duct system of *F. fessus* was examined by Dunagan and Miller (1973) who noted that the anterior portion of the uterine bell was a most dynamic structure (Figs 5G–5I) which moved eggs from the body cavity into the efferent duct. In general, the evidence in favour of egg sorting as the main function of the uterine bell seems to be stronger than the evidence against, but more comparative studies are needed and care should be taken before generalizations are made about an entire phylum when only a few species have been studied.

TABLE 5

The age structures of the egg populations in the body cavity and uterus
of Polymorphus minutus (*Palaeacanthocephala*) (After table 1, Whitfield
1970, *Parasitology* **61**, 111)

Worm age[a] (days)	Body cavity[b]		Uterus[b]	
	Mature eggs (%)	Immature eggs (%)	Mature eggs (%)	Immature eggs (%)
7	0	100	0	0
14	1	99	0	3[c]
21	11	89	100	0
28	23	77	100	0
35	32	68	100	0
56	29	71	100	0

a In each worm age group, the values in the table represent the mean for three determinations involving three female worms.

b One hundred eggs were examined from each site in each worm.

c This figure represents the total number of eggs; it is *not* a percentage value.

The uterus (Figs 5A–5F) is a muscular structure which perhaps expels eggs out of the worm on contraction. The eggs must pass through the vagina which appears to be a straight tube in some species (Figs 5A–5D) with two muscular sphincters (Dunagan and Miller, 1973). In specimens of the eoacanthocephalan *N. buttnerae*, the vagina appears to be coiled (Schmidt and Huggins, 1973). In a description of another eoacanthocephalan worm, *Acanthosentis antispinus*, Verma and Datta (1929) referred to two, single-nucleated, club-shaped glands which communicated with the vagina by a common pore close to its external aperture. They suggested that these glands might secrete some form of lubricating fluid. Further features of the terminal part of the efferent duct system are described in Section VI B.

V. GAMETOGENESIS

A. SPERMATOZOA

1. *General observations on the testis*

The testes of several species of acanthocephalan worm appear to increase in size soon after the establishment of the worms in their definitive hosts and before copulation occurs; these changes can presumably be correlated with cellular events occurring within the testis. Hardly any information was available about the general structure of the acanthocephalan testis until Whitfield (1969) and Parshad and Guraya (1979) studied the testes of the palaeacanthocephalans *P. minutus* and *C. corvi* respectively. In experimental infections of *P. minutus* in ducks, Whitfield identified a form of cytological organization typical of the early testis and one typical of the late testis. The early testis may be observed in *P. minutus* nearing the end of its development in the intermediate host and two cell types have been identified with the transmission electron microscope (Fig. 10). The first cell type or gonocyte (primordial germ cell) appears to be undifferentiated, as regular in outline, measures about 5 to 7 μm in diameter and possesses a relatively large nucleus and basophilic cytoplasm. The second type, which Whitfield assumed to be a

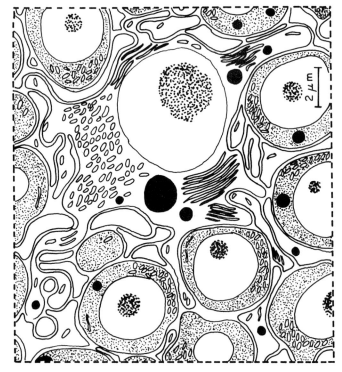

FIG. 10. Cytological interpretation of the early testis of *Polymorphus minutus*. (Redrawn from Fig. 28 of Whitfield (1969), *PhD. Dissertation, University of Cambridge*.)

supporting cell, is well differentiated, is irregular in outline, measures about 10 to 15 μm in diameter and contains a variety of cell organelles and inclusions (Fig. 10). Cytoplasmic extensions of the supporting cells were observed between the gonocytes and contacts between the cytoplasm of apparently separate supporting cells were so common that further work may reveal that the supporting tissue is syncytial in nature (Fig. 10). The late testis may be observed in a young adult *P. minutus* soon after its arrival in the small intestine of a duck. The most obvious sign of a change in cellular organization is the presence of the division products of the gonocytes which have clearly become involved in spermatogenesis (see below). The products of gonocyte division are held together by cytophores and so appear as morulae. Whitfield (1969) concluded that five spermatogonic divisions together with spermiogonic divisions occurred together with spermiogenesis in the testis of *P. minutus* within about 50 hours. The supporting tissue continues to surround and separate the germ line material. A considerable range in the timing of these testicular changes will no doubt become established as more species are studied.

Parshad and Guraya (1979) examined the general cytology of testes of *C. corvi* from naturally infected house crows, *Corvus splendens*, which were collected on four occasions during the year (February–April; May–June; August–October; November–December). Circumstantial evidence was obtained to suggest that seasonal changes may occur in the testes of *C. corvi*. On the basis of histological features, it is possible to distinguish between active, intermediate and inactive testes and in the period from May to July the majority of male worms had testes showing clear signs of spermatogenesis while those taken from November to December possessed vacuolated testes and did not appear to be supporting spermatogenesis. Interestingly, Schmidt and Kuntz (1969) had observed vacuolated tissue in the testes of *C. spilornae*. Since the course of the infection of *C. corvi* in crows has not yet been studied, it is not known whether the inactive testes belonged to very young or very old worms or whether the testes would become active in the spring when the birds breed (Ali, 1977). The difficulties involved in distinguishing between cycles of maturation and cycles of incidence, with reference to various species of Acanthocephala from fish, have been discussed by Kennedy (1975).

2. *Spermatogenesis*

Spermatogenesis is the orderly sequence of events by which the spermatogonia are transformed into spermatozoa. Three convenient phases may be recognized; spermatocytogenesis during which the spermatogonia proliferate by mitosis to form the spermatocytes; meiosis during which two divisions reduce the chromosome number by half (Table 6) and produce clusters of spermatids; spermiogenesis during which the spermatids change into the spermatozoa (Bloom and Fawcett, 1968). Although this summary is probably based largely on mammalian material, it seems to apply in general to many types of animals including the acanthocephalan species which have been studied (Table 6).

TABLE 6

Studies on spermatogenesis and spermatozoan structure in the Acanthocephala

Species	References
PALAEACANTHOCEPHALA	
Centrorhynchus milvus (b)[a]	*6, 9, 10*
Illiosentis furcatus (f)	*4, 9, 10*
Polymorphus minutus (b)	*15, 16*
Pseudoporrorchis centropi (b)	*10,*
Rhadinorhynchus pristis (f)	*9, 10*
Serrasentis socialis (f)	*8, 10*
ARCHIACANTHOCEPHALA	
Macracanthorhynchus hirudinaceus (m)	*3, 12, 13*
Mediorhynchus sp. (b)	*10*
Moniliformis cestodiformis (m)	*9, 10*
M. dubius (m)	*1, 14, 15*
Prosthenorchis sp. (m)	2
EOACANTHOCEPHALA	
Acanthosentis tilapiae (f)	*5, 9, 10*
Pallisentis golvani (f)	*7, 9, 10*
Neoechinorhynchus agilis (f)	*9, 10, 11*
Tenuisentis niloticus (f)	*9,*

[a] Definitive host; b, bird; f, fish; m, mammal.
References in italics include results from electron microscope studies.

1. Asaolu (1977)	9. Marchand and Mattei (1977b)
2. Guraya (1971)	10. Marchand and Mattei (1978a)
3. Kaiser (1893)	11. Marchand and Mattei (1978b)
4. Marchand and Mattei (1976a)	12. Meyer (1933)
5. Marchand and Mattei (1976b)	13. Robinson (1964)
6. Marchand and Mattei (1976c)	14. Robinson (1965)
7. Marchand and Mattei (1976d)	15. Whitfield (1971a)
8. Marchand and Mattei (1977a)	16. Whitfield (1971b)

Most work on acanthocephalan spermatogenesis to date has been concerned with spermiogenesis (Kaiser, 1893; Whitfield, 1971a; Marchand and Mattei, 1976a, 1977a, 1978a, b). However, the work of these authors and that of Robinson (1964, 1965) on *Mac. hirudinaceus* and *M. dubius* and Guraya (1971) on *Prosthenorchis* sp. indicate that normal spermatocytogenesis and meiosis occur before the spermatids are formed. Spermiogenesis in *P. minutus* has been described by following with the electron microscope some of the events which happen to a morula of spermatids in the late testis (see above). Each morula is a multinucleate syncytium consisting of as many as 64 spermatids (Whitfield, 1971a). Each early spermatid appears to contain a relatively large nucleus, concentrations of ribosomes, little or no obvious rough endoplasmic reticulum, a few mitochondria and a centriole (Fig. 11A). As development proceeds, membrane-bounded, electron-dense inclusions are formed in the cytoplasm of each spermatid and a prominent Golgi body becomes apparent (Fig. 11B). The development of the flagellum is the next major event to occur and this involves the centriole. The flagellum, with its characteristic complement of microtubules (9+2) extends out of the cell

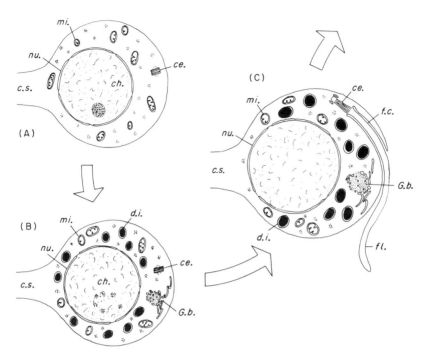

Fig. 11. Diagrammatic interpretation of events occurring during spermiogenesis in the palaeacanthocephalan *Polymorphus minutus*. (Redrawn from Figs 1, 2 and 5, Whitfield (1971a), *Parasitology* **62**, 415.) (A) Early preflagellar spermatid. (B) Late preflagellar spermatid. (C) Development of the flagellum. For details of abbreviations, see p. 85.

body of the spermatid roughly opposite to the cytophore which connects the cell body to the morula (Fig. 11C). The flagellum at first passes through the cytoplasm of the spermatid in a deep cleft (Fig. 11C), but later the cleft and its enclosed flagellum are observed to have become located in a deep invagination in the surface of the nucleus. The nucleus now becomes very elongate and most of the nuclear envelope disintegrates to form the spermatozoan body which contains the chromatin and a series of electron dense inclusions along both sides of the flagellum (see Fig. 12). There was no direct evidence of an acrosome in the fully developed spermatozoon of *P. minutus* and no sign of a mitochondrion, although the spermatozoa are known to be motile both *in vivo* and *in vitro* (Whitfield, 1971a; D. W. T. Crompton, unpublished observations). The mitochondria appear to be left in the cytoplasm of the cytophore which remains when the spermatozoon becomes a free cell. Throughout this study, Whitfield, who appears to have been the first worker to study acanthocephalan spermiogenesis with the electron microscope, was aware that his interpretation of events occurring in *P. minutus* were based on a series of static images of a continuously changing living process.

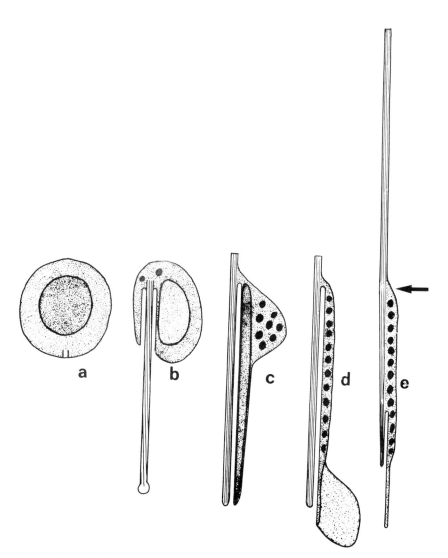

FIG. 12. The main phases of spermiogenesis in the eoacanthocephalan *Neoechinorhynchus agilis*. (Redrawn from Fig. 1 of Marchand and Mattei (1978b), *Journal of Ultrastructure Research* **63,** 41.) (a) Young rounded spermatid in which the centriole is applied against the plasma membrane. (b) Flagellar growth posteriorly in which the outer microtubules of the axoneme are arranged helically. (c) An older spermatid in which the nucleus has become elongated and the flagellum has started to extend anteriorly. (d) A spermatid near the end of spermiogenesis in which elimination of the cytoplasmic drop and the formation of the nucleocytoplasmic derivative (=spermatozoan body) are taking place. (e) Mature spermatozoon; the main part of the flagellum extends anteriorly.

The researches of Marchand and Mattei have involved spermiogenesis and spermatozoa in 11 species of Acanthocephala from the three orders of the phylum (Table 6). Their interpretation of spermiogenesis in all these species is summarized diagrammatically for *N. agilis* in Fig. 12. At the start of spermiogenesis, the centriole migrates to the anterior part of the young spermatid and adopts a position beyond the anterior extremity of the nucleus. The flagellum begins to grow posteriorly for a while and the nucleus then changes in shape and extends beyond the end of the flagellum. The nuclear envelope breaks down, glycogen appears in the cytoplasm, proteinaceous granules are formed, chromatin condensation occurs and eventually the nucleocytoplasmic derivative (=spermatozoan body of Whitfield, 1971a, b) is formed (Fig. 12). During the formation of the nuclear cytoplasmic derivative, a drop of cytoplasm is eliminated from the spermatid. The main growth of the flagellum occurs anteriorly to complete the formation of the spermatozoon which now appears to show reversed anatomy. The observations and interpretations of Marchand and Mattei differ in various details from those made by Whitfield (1971a) on *P. minutus*, particularly with regard to the anteriorly directed growth of the flagellum. The species studied by Marchand and Mattei (Table 6), like *P. minutus*, do not appear to contain mitochondria, but these are eliminated in the cytoplasmic drop (Marchand and Mattei, 1978a) rather than remaining in the cytophore (Fig. 11c; Whitfield, 1971a).

3. *Structure of the spermatozoa*

The mature spermatozoa of most of the species of Acanthocephala that have been studied to date (Table 6) are basically filiform cells and seem to

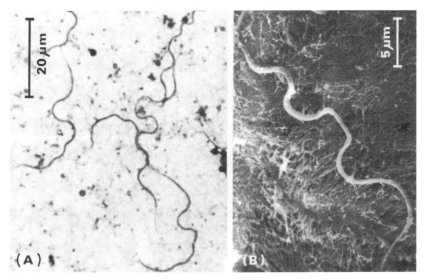

Fɪɢ. 13. Spermatozoa of *Moniliformis dubius* (Archiacanthocephala). (A) Photomicrograph of fixed, Giemsa-stained, mature spermatozoa from a 70-day-old male. (B) Scanning electron micrograph of a spermatozoon on the surface of an ovarian ball from the body cavity of an inseminated female.

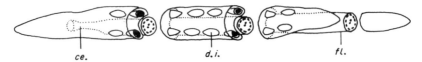

FIG. 14. Simplified cutaway stereogram of a spermatozoon of the palaeacanthocephalan *Polymorphus minutus*. The drawing omits large portions of the cell in the cut regions. (Redrawn from Fig. 7 of Whitfield (1971a), *Parasitology* **62**, 415.) For details of abbreviations, see p. 85.

measure from about 25–65 μm in length and about 0·2–0·5 μm in width (Figs 13 and 14). In effect, there is no obvious differentiation into a head, middle piece or tail (Austin, 1965). The spermatozoa of the palaeacantho-cephalan *Illiosentis furcatus* (Marchand and Mattei, 1976a) appear to be unusual in that a distinct ovoid body is clearly recognizable at one end. Guraya (1971) referred to this as the sperm head, but perhaps some other term should be used until the anterior end of the spermatozoon has been identified and the nuclear material located (see above).

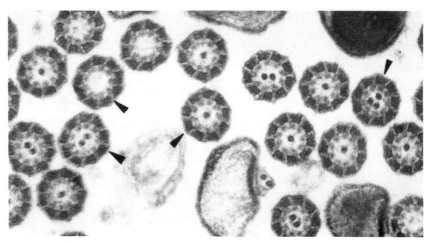

FIG. 15. Transmission electron micrograph showing transverse sections through mature spermatozoa in the vesicula seminalis of the eoacanthocephalan *Acanthosentis tilapiae*. The arrows point to various microtubular arrangements within the flagella (see Table 7). (After Fig. 4 of Marchand and Mattei (1976b), *Journal of Ultrastructure Research* **55**, 391.)

Electron micrographs published in various papers cited in Table 6 reveal that the flagellum contains a fairly typical axoneme consisting of micro-tubules. The usual combination of nine double outer microtubules and a pair of central microtubules is present in the flagella of most species; in some micrographs (Fig. 15) evidence has been obtained to indicate the exis-tence of radial links (Warner, 1974). In *N. agilis, Pallisentis golvani, I. furcatus, Rhadinorhynchus pristis*, and probably in *Serrasentis socialis*, the A microtubule only is present in the outer circle of that part of the axoneme

TABLE 7

Observations on the flagella of acanthocephalan spermatozoa[a]

Species	TS[b]	Mean percentages (±S.E.) of microtubule patterns					
		9+0	9+1	9+2	9+3	9+4	9+5
Eo[c]. *Acanthosentis tilapiae* (12)[d]	9886	7·31±9·54	18·47±46·23	73·23±23·80	0·97±0·89	—	—
Eo. *Pallisentis golvani* (2)	840	7·26±1.16	2·68±01·82	88·46±00·98	1·40±1·46	0·17±0·24	—
Eo. *Neoechinorhynchus agilis* (19)	9377	0·04±0·03	0·15±0·07	2·22±00·57	95·66±1·04	1·90±0·83	0·09[e]
Eo. *Tenuisentis niloticus* (1)	93	1·07	1·07	91·39	5·37	1·07	—

[a] All information in the table and footnotes is from Marchand and Mattei (1977b) unless stated otherwise.
[b] Number of transverse sections of spermatozoa examined in the vesicula seminalis.
[c] Eo., Eoacanthocephala; Ar., Archiacanthocephala; Pa., Palaeacanthocephala.
[d] Number of worms involved.
[e] From a sample of 1059 sections from one worm.

The following qualitative observations have also been published; Pa., *Polymorphus minutus*, 9+2 (Whitfield, 1971a); *Illiosentis furcatus*, usually 9+0 (Marchand and Mattei, 1976a) occasionally 9+1; *Centrorhynchus milvus*, usually 9+2 (Marchand and Mattei, 1976c) occasionally 9+3; *Rhadinorhynchus pristis*, usually 9+2 (Marchand and Mattei, 1976b) occasionally 9+3; *Serrasentis socialis*, usually 9+2 occasionally 9+3; Ar., *Moniliformis dubius*, 9+3 (Whitfield, 1971a); *M. cestodiformis*, usually 9+2 occasionally 9+3.

which is associated with the nucleocytoplasmic derivative (Fig. 12, Marchand and Mattei, 1978b). Considerable variations have been observed in the numbers of central microtubules both within and between species of Acanthocephala (Table 7; Fig. 15). The functional, taxonomic and phylogenetic significance of this variability is not understood. Live spermatozoa normally display propagated sine waves in the same direction along their length. It is to be expected that this motility will eventually be shown to be associated with ATPase activity in the arms of the A microtubules in the outer circle of nine microtubules (see Whitfield, 1971a).

None of the studies carried out so far have identified an acrosome in acanthocephalan spermatozoa (Table 6). Guraya (1971) and Whitfield (1971a) concluded that the acrosome was absent from the spermatozoa of *Prosthenorchis* sp. and *P. minutus* respectively. During spermiogenesis in animals, the Golgi body is known to contribute to the formation of the acrosome which subsequently facilitates fertilization (Austin, 1965). It is interesting to note that, although no typical acrosome can be recognized in the mature spermatozoon of *S. socialis*, a centriolar derivative, which is formed in close association with the Golgi body, may be involved in some aspects of fertilization (Marchand and Mattei, 1977a).

B. OOCYTES

1. *The ovarian ball*

Earlier workers, especially Meyer, recognized the functional significance of the ovarian ball after studying its structure and cytology by means of the light microscope (see Hamann, 1891; Kaiser, 1893; Meyer, 1928, 1933; Nicholas and Hynes, 1963). These authors were convinced that the oocytes developed from germ-line syncytial tissue in the middle of the ovarian ball (Fig. 16A). A more detailed understanding of the ovarian ball from representatives of each of the acanthocephalan orders became available as a result of studies with the transmission electron microscope (Table 8). In ovarian balls from mature female *P. minutus* (Crompton and Whitfield, 1974), *M. dubius* (Crompton and Whitfield, 1974; Atkinson and Byram, 1976) and *Aca. tilapiae* and *Pal. golvani* (Marchand and Mattei, 1976d), three components may be recognized; these are considered to be two separate multinucleate syncytia and a cellular zone (Fig. 16B). The inner region consists mainly of the oogonial syncytium which appears to give rise to germ cells and the other elements of the cellular zone. The cells and the oogonial syncytium are embedded in the supporting system which also forms the boundary of the ovarian ball (Fig. 16B). The elaborate surface morphology of the ovarian ball (Figs 8B and 8C) and ultrastructural (Crompton and Whitfield, 1974; Atkinson and Byram, 1976) and histochemical (Parshad and Guraya, 1977b) observations suggest that the supporting syncytium has an important nutritive function in the ovarian ball. This brief description of the structure of an ovarian ball does not give any indication of either its complexity (Fig. 17) or its internal organization. In *C. corvi*, the mature ovarian ball, as seen with the light microscope (Fig. 8F), appears to be composed of about 24 to 30

TABLE 8

Studies on ovarian ball structure and oogenesis in the Acanthocephala

Species	References
PALAEACANTHOCEPHALA	
Centrorhynchus corvi (b)[a]	13, 14
Leptorhynchoides thecatus (f)	6
Polymorphus minutus (b)	7, 12
Pomphorhynchus laevis (f)	*18*
ARCHIACANTHOCEPHALA	
Macracanthorhynchus hirudinaceus (m)	10, 11, 16
Moniliformis dubius (m)	*2, 3, 4*, 7, *15*, 17
Prosthenorchis sp. (m)	8
EOACANTHOCEPHALA	
Acanthosentis oligospinus (f)	1
A. tilapiae (f)	*9*
Neoechinorhynchus cylindratus (f)	5
Pallisentis golvani (f)	*9*

[a] Definitive host; b, bird; f, fish; m, mammal.
References in italics include results from electron microscope studies.

1. Anantaraman and Subramoniam (1975)
2. Asaolu (1976)
3. Asaolu (1977)
4. Atkinson and Byram (1976)
5. Bone (1974a)
6. Bone (1974b)
7. Crompton and Whitfield (1974)
8. Guraya (1969)
9. Marchand and Mattei (1976d)
10. Meyer (1928)
11. Meyer (1933)
12. Nicholas and Hynes (1963)
13. Parshad and Guraya (1977b)
14. Parshad and Guraya (1978)
15. Parshad *et al.* (1980b)
16. Robinson (1964)
17. Robinson (1965)
18. Stranack (1972)

functional units, each of which consists of a portion of oogonial syncytium surrounded by various cells (Parshad and Guraya, 1977b). The units are arranged in a regular pattern and are embedded by supporting syncytium which may facilitate some form of contact between them.

The two syncytia in the mature ovarian ball (Fig. 16B) may be distinguished by their appearance in the electron microscope. The oogonial cytoplasm contains aggregations of rough endoplasmic reticulum, characteristic mitochondria and electron-dense, spherical inclusions which increase in number and can be traced as the oocytes develop (Crompton and Whitfield, 1974). These cytoplasmic 'markers' serve to confirm that the oogonial syncytium is correctly named. The cytoplasm of the supporting syncytium also contains characteristic elongate mitochondria, but no dense inclusions. In addition, the cortical region of the supporting syncytium of *M. dubius* appears to possess many microfilaments (Crompton and Whitfield, 1974). The origins of the two syncytia are not yet well understood although a mechanism for their formation in *M. dubius* has been proposed by Atkinson and Byram (1976). According to these authors, 7–9 days after the establishment of *M. dubius* in the rat, the genital primordium is transformed into discrete ovarian balls when free germinal cells become enveloped by somatic tissue which originates from what they identified as the ligament sac primordium. This

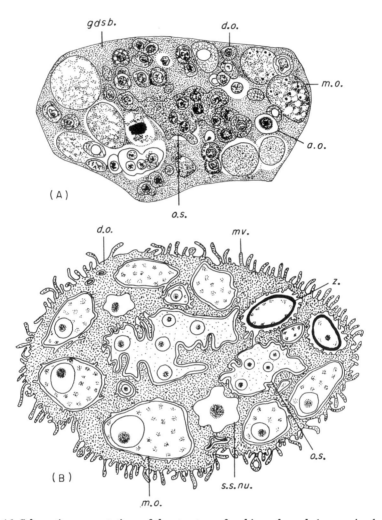

FIG. 16. Schematic representations of the structure of archiacanthocephalan ovarian balls. (A) *Macracanthorhynchus hirudinaceus* as observed with the light microscope. (Redrawn from Fig. 345 of Meyer (1933), *Acanthocephala*, "Bronns *Klassen und Ordnungen des Tierreichs*".) (B) *Moniliformis dubius* as observed with the transmission electron microscope. (Redrawn from Fig. 28 of Atkinson and Byram (1976), *Journal of Morphology* **148**, 391.) The terminology is that proposed by Crompton and Whitfield (1974). For details of abbreviations, see p. 85.

view is not easy to accept for several reasons, particularly because it involves the timing of ovarian ball formation. First, ovarian balls may readily be observed and counted in *M. dubius* which are less than 7-days-old (D. W. T. Crompton and S. Arnold, unpublished observations). This facility is the basis of the identification of young female worms (Crompton, 1974; Crompton *et al.*, 1976). Secondly, Yamaguti and Miayata (1942), Moore (1946) and Asaolu (1976) observed immature ovarian balls in *M. dubius* in their intermediate hosts. These observations do not mean that the two syncytia had been formed before female *M. dubius* had been established in the rat for a week. They do imply, however, that any intermingling of somatic and germinal components could easily have occurred relatively early in development. Thirdly, Asaolu (1976, 1977) made an ultrastructural study of immature ovarian balls in *M. dubius* from cockroaches and demonstrated the existence of a level of cytological organization equally complex as that depicted by Atkinson and Byram (1976) in $7\frac{1}{2}$-day-old adult worms from rats. Fourthly, it is difficult to make dynamic interpretations about cell movements from electron micrographs of fixed tissue. How does the microscopist know that fixed tissue is "extending a long tendril of cytoplasm to envelop a primordial germ cell" (Atkinson and Byram, 1976)? Further work should be undertaken to confirm Atkinson and Byram's view and to explore the possibility that special embryonic cytoplasmic factors may have a role in the determination of the germ cells (see Davenport, 1979). The earlier during development that the germ line can be identified, the better will be our understanding of the origin of the ovarian balls and their component syncytia.

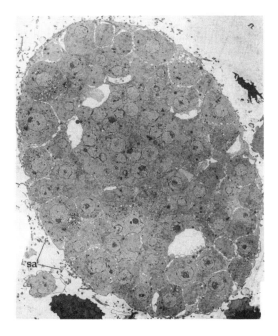

FIG. 17. Transmission electron micrograph through an ovarian ball from an inseminated *Acanthosentis tilapiae* (Eoacanthocephala). (After Fig. 1 of Marchand and Mattei (1976d), *Journal of Ultrastructure Research* **56**, 331.) For details of abbreviation, see p. 85.

2. *Oogenesis*

The available evidence (Table 8) suggests that oogenesis in the Acantho-cephala follows the expected pattern involving, successively, the oogenia, primary oocytes, secondary oocytes and mature oocytes or egg cells (Daven-port, 1979). The reduction division probably occurs when the primary oocytes divide, and growth, associated with the accumulation of reserve substances, probably happens while the secondary oocyte matures. The formation of the oogonia is believed to result from the detachment of small uninucleate portions of cytoplasm from the oogonial syncytium (Fig. 16B; Meyer, 1933; Crompton and Whitfield, 1974; Atkinson and Byram, 1976; Parshad and Guraya, 1977b). Three types of nuclei have been identified in the oogonial syncytium of *C. corvi* and the largest ones are those that appear in the free oogonia (Parshad and Guraya, 1977b). These independent cells now form part of the cellular zone (Fig. 16B) and as far as is known, the surrounding syncytial tissue keeps them in physical isolation from all other cells and from the oogonial syncytium (Crompton and Whitfield, 1974; Fig. 16B). The free oogonia, after further division, are next assumed to become primary oocytes and Atkinson and Byram (1976) obtained some evidence from cells in ovarian balls from 12-day-old *M. dubius* to suggest that meiotic prophase was occur-ring. Similarly, Parshad and Guraya (1977b) concluded that many of the growing primary oocytes of *C. corvi* were undergoing meiotic prophase when the ovarian balls were fixed. This condition supports the view that oogenesis in the Acanthocephala is generally similar to that observed in other animals (Davenport, 1979).

Further evidence confirms that the developing oocytes grow and become centres of metabolic activity. The cytoplasm of the developing oocytes of *M. dubius* and *P. minutus* is seen to contain Golgi bodies, rough endoplasmic reticulum, annulate lamellae and membrane-bounded, electron-dense spherical inclusions (Crompton and Whitfield, 1974; Atkinson and Byram, 1976). This description would appear to apply to the eoacanthocephalans *Aca. tilapiae* (Fig. 17) and *Pal. golvani* (Marchand and Mattei, 1976d). Histochemical tests indicate that lipids, phospholipids, lipoproteins, proteins and RNA accumulate during the growth phase of the oocytes *Prosthenorchis* sp. (Guraya, 1969), *Aca. oligospinus* (Anantaraman and Subramoniam, 1975) and *C. corvi* (Parshad and Guraya, 1977b). The mature oocytes are relatively large cells which appear to contain many cytoplasmic inclusions and are usually located just beneath the cortical region of the supporting syncytium near the surface of the ovarian ball (Fig. 16B).

3. *Oocyte atresia*

Theoretically, the simplest concept of the functioning of the mature ovarian ball might predict a steady production of oogonia from the oogonial syncytium in balance with a steady release of zygotes from the surface into the body cavity (see Section VIII). In practice, all the mature oocytes do not neces-sarily become fertilized, and Meyer (1928) described and figured (Fig. 16A) the presence of "abortive Keimzellen" (= ? degenerating germ cells in the

ovarian balls of *Mac. hirudinaceus*). Autolysis and degeneration of some
mature oocytes were considered to occur in ovarian balls from uninseminated *M. dubius* (Crompton and Whitfield, 1974). Similar observations on
oocyte atresia involving growing and mature oocytes were made by Parshad
and Guraya (1978) from both uninseminated and inseminated *C. corvi*.
On the basis of counting the numbers of oocytes visible in tissue sections of *C. corvi*, about 8·5% of the growing oocytes from inseminated worms
and 32% from uninseminated worms were considered to be in some state
of degeneration. The implications of this observation are that fertilization,
which would lead to the loss of cells from the ovarian ball, has some effect
on the rate of oocyte atresia and that the pressure to produce oogonia from
the oogonial syncytium (Fig. 16B) may be greater than the pressure to complete
oogenesis or retain mature oocytes. Oocyte atresia, however, is a widespread
and still unexplained process in many animal species which have been
investigated in much greater detail than the Acanthocephala (Cohen, 1977).
Parshad and Guraya (1978) suggested that products of the degenerating
oocytes probably became incorporated into the cytoplasm of the supporting
syncytium. In contrast to the above observations are those of Anantaraman
and Subramoniam (1975) who identified free oocytes, which were thought to
be undergoing resorption, in the fluid of the body cavity of *Aca. oligospinus*.

VI. Mating

A. Sexual Congress

Sexual congress, which has been defined as the association of males and
females for sexually reproductive purposes (Cohen, 1977), is very difficult to
observe or study in the case of the endoparasitic Acanthocephala. Probably
most information about mating behaviour may be deduced for *M. dubius*
from the results of several experimental studies of its course of infection in
the rat (Burlingame and Chandler, 1941; Holmes, 1961; Crompton *et al.*,
1972; Crompton, 1974, 1975; Abele and Gilchrist, 1977; Nesheim *et al.*,
1977, 1978; Parshad *et al.*, 1980a).

Moniliformis dubius and the adults of other species occupy fairly precise
sites in the small intestines of their definitive hosts (Crompton, 1975) although
the location of these sites may change during the course of the infection and
the maturation of the worms (Awachie, 1966). Newly established *M. dubius*
tend to be located well into the posterior half of the small intestine of the rat
(Crompton, 1975; Nesheim *et al.*, 1977). About 3 or 4 weeks later, most
M. dubius are observed in a compact group attached to the mucosa in the
anterior half of the small intestine (see above references); this anterior emigration of about 50 cm in an adult rat seems to be part of a behaviour pattern
which is most obvious when the host's diet has the capacity to support the
growth and reproduction of the worm (Nesheim *et al.*, 1977, 1978; Parshad
et al., 1980a). The apparent timing of the emigration coincides with the period
when, on the basis of the occurrence of fertilization (Crompton, 1974;
Atkinson and Byram, 1976) and the beginning of egg release (Burlingame

TABLE 9

Estimates of the timing of insemination, egg formation and egg release by Acanthocephala

Species	Pre-patent period		Start of egg release	Patent period (egg release)	Estimate of mean daily egg release /female	References
	Phase 1 (first insemination)	Phase 2 (egg formation)				
Ar.[a] *Moniliformis dubius* (m)[b]	16 days	~22 days	38 days	106±16 days	~5500	1, 2, 4, 5, 6, 7, 8, 13, 15, 16
Pa. *Polymorphus minutus* (b)	5 days	~17 days	22 days	~21 days	~1700	6, 9, 10, 11, 12, 14
Pa. *Echinorhynchus truttae* (f)	3 days	~8 weeks	64 days	3 weeks	—	3
Pa. *Leptorhynchoides thecatus* (f)	3–4 weeks	~4 weeks	~8 weeks	—	—	8
Ar. *Macracanthorhynchus hirudinaceus* (m)	—	—	~10 weeks	~10 months	~260,000	9
Ar. *Mediorhynchus centurorum* (b)	—	—	35 days	84 days	—	10

[a] Ar., Archiacanthocephala; Pa., Palaeacanthocephala.   [b] Definitive host: b, bird; f, fish; m, mammal.

References:
1. Abele and Gilchrist (1977)
2. Atkinson and Byram (1976)
3. Awachie (1966)
4. Burlingame and Chandler (1941)
5. Crompton (1974)
6. Crompton and Whitfield (1968)
7. Crompton et al. (1976)
8. DeGiusti (1949)
9. Kates (1944)
10. Nickol (1977)
11. Nicholas and Hynes (1958)
12. Petrochenko (1958)
13. Robinson (1965)
14. Romanovski (1964)
15. Sita (1949)
16. Yamaguti and Miyata (1942)

and Chandler, 1941; Crompton *et al.*, 1972; Table 9) copulation would be expected to occur frequently. Crompton (1975) presented the results of a preliminary analysis of the distribution of *M. dubius* of known age in rats fed on a diet of high starch content (Nesheim *et al.*, 1977). Worms aged 4 weeks were found close together in the anterior half of the small intestine and it was possible to claim that each female *M. dubius* was, on average, in contact with the maximum number of males and also that the highest percentage of the females present in the rat were experiencing contact with at least one male. This is also the period of infection when the sex ratio would be 1 (Section III). Worms aged 8 weeks were found more posteriorly in the small intestine and appeared to have fewer physical contacts with other worms. It is interesting to note that the time when *M. dubius* was considered to have maximum contact was within the period for which males and females must be together to ensure a long period of egg production (Crompton *et al.*, 1972; Crompton, 1974).

The observation that the start of egg production by *M. dubius* is not disturbed by the earlier surgical transplantation of worms of less than 7 days old from one rat to another indicates that the worms come into contact with each other although they have not begun their association in the rat naturally (Crompton, 1974). Similarly, the findings that 1 male *M. dubius* is capable of inseminating as many as 17 out of 27 females of the same age and that young males can inseminate older females following transplantation (Crompton, 1974) may imply that some form of sexual recognition exists (Section II). Care must be taken before any of these observations on experimental infections of *M. dubius* are applied to either more complicated natural infections or to other species of Acanthocephala.

B.   COPULATION AND INSEMINATION

Further information about copulation and the copulatory apparatus is given in Table 9 and Fig. 18 in addition to that included in Section IV B. Much more comparative information is needed about when copulation first begins in different species of Acanthocephala (Table 9). DeGiusti (1949) concluded that copulation and fertilization probably did not occur in *L. thecatus* until the worms had been established in their fish hosts for at least 3 weeks. In contrast, Brattey (1980) has observed active spermatozoa in cystacanths of *A. lucii* which is another palaeacanthocephalan from fish definitive hosts. Copulation between male and female *A. lucii* occurred within 24h of the experimental infection of perch (*Perca fluviatilis*) which were kept at 20°C in the laboratory (Brattey, 1980). The timing of events in Table 9 are approximations based on laboratory observations.

It seems likely that the male acanthocephalan may have the more active role in copulation than the female. The relevant anatomy of the male (Figs 3A–3E, 4G, 4I, 18B, 18D–18G) is more complex than that of the female (Figs 5A–5D, 6A–6F) and the nervous system of the male (Fig. 18E) is more extensive. Possession of the greater neural capacity by male Acanthocephala also suggests that they may be more active during sexual congress than the

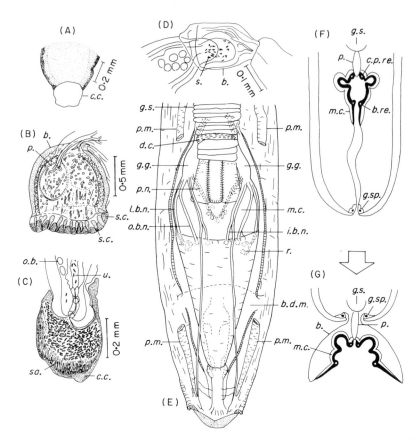

FIG. 18. Acanthocephalan copulation. (A) Sketch of a copulation cap attached to the posterior end of a mature female *Corynosoma constrictum* (Pa.). (Redrawn from Fig. 11 of Van Cleave (1920), *Transactions of the Illinois Academy of Science* 13, 280.) (B) Features of the copulatory bursa of *Fessisentis fessus* (Pa.). (Redrawn from Fig. 2B of Dunagan and Miller (1973), *Proceedings of the Helminthological Society of Washington* **40**, 209.) (C) Sagittal section through the copulation cap and posterior end of a female *Neoechinorhynchus emyditoides* (Eo.). (Redrawn from Fig. 22 of Fisher (1960), *Journal of Parasitology* **46**, 257). (D) Posterior extremities of a male and female *Gorgorhynchus clavatus* (Pa.) fixed *in copula*. (Redrawn from Pl. 54 and Fig. 16 of Van Cleave (1940), *Allen Hancock Foundation: Publications*, Series 1, vol. 2, 501.) (E) Diagrammatic interpretation of the main nerves in the copulatory bursal region of *Macracanthorhynchus hirudinaceus* (Ar.). (Redrawn from Fig. 1 of Dunagan and Miller (1979), *Proceedings of the Helminthological Society of Washington* **46**, 106.) (F and G) Diagrams illustrating some morphological rearrangements which occur when the inverted (F) copulatory bursa of *Polymorphus minutus* (Pa.) becomes everted (G). (Redrawn from Figs. 26 and 27 of Whitfield (1969), *PhD. Dissertation, University of Cambridge*.) Ar., Archiacanthocephala; Eo., Eoacanthocephala, Pa., Palaeacanthocephala. For details of other abbreviations, see p. 85.

females. Until recently, it was known that both male and female Acanthocephala possessed a cerebral ganglion and that males also possessed a genital
ganglion (or paired ganglia) consisting of two groups of cells connected by at
least one commissure (Meyer, 1933; Figs 4B, 4F, 4G, 4H, 18E), but a third
ganglion, the bursal ganglion has now been described from male *M. moniliformis* and *Mac. hirudinaceus* (Dunagan and Miller, 1973, 1977, 1979). The
genital ganglia of *Mac. hirudinaceus* measure about 675 μm in length and
contain, perhaps rather surprisingly since acanthocephalan organs are often
considered to be eutelic (Van Cleave, 1914), a variable number of cells
(Dunagan and Miller, 1979). Sensory and motor neurons are present and
there are nerve connections with Saefftigen's pouch, the surrounding muscles,
the penis and the bursa; at least eight receptors have been found in the bursal
musculature (Fig. 18E). The bursal ganglion of male *M. moniliformis* consists
of four large cells which are arranged in two pairs against the dorsal longitudinal muscles of the body (Dunagan and Miller, 1977). The ganglion is located
about 1 mm from the posterior end of a male with an inverted bursa. Dunagan
and Miller suggest that the activities of the bursal ganglion may be coordinated
with those of the genital ganglion through nerves that innervate the muscles
responsible for bursal eversion and withdrawal. Other delicate muscular
movements of the bursa and penis will also be involved in copulation if
spermatozoa are to be transferred successfully, particularly if the location of
the female genital pore is as variable in other species as may be the case in
F. fessus (Nickol, 1972).

Only a few species of Acanthocephala have been observed *in copula* and
these include *Sphaerechinorhynchus rotundocapitatus* (Johnston and Deland,
1929), *Aca. dattai* (Podder, 1938), *Gorgorhynchus clavatus* (Fig. 18D; Van
Cleave, 1940), *Pseudoporrorchis rotundatus* (Golvan, 1956), *Acanthogyrus
partispinus* (Furtado, 1963) and *M. dubius* (Atkinson and Byram, 1976).
Podder (1938) stated that during copulation the everted bursa of the male
became attached to the posterior end of the female and that by a simple
sucking mechanism the vagina of the female was drawn into the bursal cavity
so that the penis could be inserted. This oversimplified view of copulation
nevertheless emphasizes the need for relatively elaborate neuromuscular
coordination (Figs 18F and 18G), and behaviour. *Neoechinorhynchus* spp. was
observed to copulate *in vitro* by Dunagan (1962) and an extension of his
work could lead to a better understanding of this aspect of acanthocephalan
reproduction.

The posterior part of the body in some species of Acanthocephala is
endowed with small spines (Fig. 18D) which are usually called the genital
spines (Van Cleave, 1920). The spines on the male are unlikely to have any
copulatory function but those in the female may become embedded in the
bursal tissue and so strengthen the copulatory union. When copulation has
occurred, the posterior end of the female is often covered by a copulatory
cap (Figs 18A and 18C) which is believed to be produced by the secretions of
the cement glands (Section IV B). The presence of the cap is a useful, but not
entirely reliable, indicator of insemination which is best determined by

examining the ovarian balls for the presence of zygotes (Crompton, 1974). It has been generally assumed that the function of the cement is to assist in holding the partners together during copulation and that the residual cap prevents the loss of spermatozoa from the female (Hyman, 1951; Crompton, 1970). However, if the caps are not lost, egg release may be impeded and subsequent insemination prevented, assuming that more than one insemination is necessary if a full patent period is to be experienced by the female. Fisher (1960) cut sections through a female *N. emyditoides* and found both eggs and spermatozoa within what appeared to be a chamber within the cap (Fig. 18C). This most interesting and isolated observation suggests that sometimes the cap may serve as some kind of storage vessel for spermatozoa.

A stimulating view about the significance of the copulation cap has recently been proposed by Abele and Gilchrist (1977). They noted from their own and published observations that copulation caps occur on males of *M. dubius*, *A. parksidei*, *E. truttae* and *P. minutus* and they interpreted this finding as an example of homosexual rape which would remove temporarily some of the males from the reproductive population. Apparently, only cement and not spermatozoa were transferred to male *M. dubius*. This observation strongly supports Abele and Gilchrist's hypothesis and indicates that the males are able to distinguish between males and females. Earlier, Awachie (1966) suggested that the copulation caps on male *E. truttae* might indicate that sex recognition was poor in this species.

Very little is known about insemination in the Acanthocephala. In the majority of animal species, the spermatozoa are transferred to the female in a fluid medium known as the semen which is secreted by a variety of glands (Cohen, 1977). Presumably acanthocephalan spermatozoa are also transferred in some form of fluid. In other animal species, the spermatozoa are enclosed in a spermatophore before being accepted by the female (Austin, 1965). There is no positive reference in the literature to spermatophore production by an acanthocephalan (Meyer, 1933) although Atkinson and Byram (1976) ventured to suggest that spherical objects measuring about 55 µm in diameter and situated in the body cavities of 18-, 19- and 154-day-old *M. dubius* could have been spermatophores.

Acanthocephalan spermatozoa are mobile when observed *in vitro* in physiological saline after removal from the male or from the fluid in the body cavity of a female (Marchand and Mattei, 1978a; D. W. T. Crompton and P. J. Whitfield, unpublished observations). Studies with the electron microscope reveal that on entering the female, the spermatozoa of 10 species undergo certain changes. Glycogen, formed during spermiogenesis, is no longer obvious in the cytoplasm of the sperm body and various other inclusions also disappear (Marchand and Mattei, 1978a). Spermatozoa, however, may be observed inside female *M. dubius* from rats that no longer contain males (Atkinson and Byram, 1976). It is not known whether the spermatozoa are stored in any particular organ of the female, whether they are sustained by nutritive properties of the body cavity fluid or whether they can survive for some time in association with the ovarian ball (Fig. 13B). Spermatozoa have been observed in contact with the supporting syncytium at the surface of the ovarian balls of representative species of the three acanthocephalan orders (Meyer, 1928, 1933; Baer, 1961; Nicholas and Hynes, 1963; Crompton and Whitfield, 1974; Atkinson and Byram, 1976; Marchand and Mattei, 1976d). Until information is available about spermatozoan storage, spermatozoan wastage (Cohen, 1977), and the average number of spermatozoa transferred during insemination, there is little point in speculating in too much detail about the number of times insemination must occur if a female worm is to achieve maximum egg production. Crompton (1974) demonstrated under experimental conditions that the period of egg release by female *M. dubius* was curtailed in cases where the females were isolated from the males earlier than 5 weeks after the start of the infection. Since copulation and insemination can occur when *M. dubius* are 15–16 days old (Crompton, 1974; Atkinson and Byram, 1976), it could be argued that the female may need to mate several times in order to acquire sufficient spermatozoa for full fecundity. This type of reasoning may not apply to other acanthocephalan species.

VII. Fertilization

Van Cleave (1953) concluded that fertilization took place when the ova (mature oocytes) became separated from the ovarian balls (Figs 16A and 16B) and this view seemed to be shared by Guraya (1969) and by Nicholas and Hynes (1963) despite the fact that Meyer (1928, 1933) had already demonstrated that spermatozoa became attached to the surfaces of the ovarian balls (see Figs 13B and 17). Nicholas and Hynes (1963) also drew attention to small spindle-shaped, Feulgen-positive bodies inside the ovarian balls of *P. minutus;* they identified these bodies as spermatozoan heads. Anantaraman and Subramoniam (1975) observed mature oocytes in the body cavity of *Aca. oligospinus* but suggested that these cells might have been undergoing resorption. Marchand and Mattei (1976d) noticed free mature oocytes of *Pal. golvani* on two occasions and considered that accidental rupture of the supporting syncytium must have occurred. All the recent evidence, however, supports the view that in *P. minutus* (Crompton and Whitfield, 1974), *M. dubius* (Crompton and Whitfield, 1974; Atkinson and Byram, 1976), *Aca. tilapiae* and *Pal. golvani* (Marchand and Mattei, 1976d), which are representatives of the three acanthocephalan orders, fertilization either occurs or begins while the mature oocytes are contained within the ovarian ball.

Fertilization involves several sequential processes rather than a precise action. Although these may differ from animal to animal, there seem to be basically five which should be expected (Cohen, 1977). These are: (1) penetration of the oocyte envelopes by the spermatozoan; (2) some form of membrane fusion between the spermatozoan surface and the oocyte so that the male and female cytoplasm are brought into contact; (3) the entry of at least the spermatozoan nucleus and a centriole, and even the axonome, into the oocyte; (4) the activation of the egg which often includes a prominent cortical reaction resulting in a block to polyspermy, metabolic and synthetic changes and the formation of polar bodies; (5) syngamy or the fusion of nuclear materials.

Acanthocephalan spermatozoa must first pass through the supporting syncytium (Figs 16B and 17) before contact can take place with a mature oocyte. This process, which is clearly an extra aspect of fertilization that may be peculiar to the Acanthocephala (see above), has not yet been explained, although Atkinson and Byram (1976) have interpreted a transmission electron micrograph of a spermatozoon at the surface of an ovarian ball of *M. dubius* as showing possible fusion between the spermatozoan plasmalemma and the microvilli (Fig. 16B) at the surface of the supporting syncytium. There are also various channels and relatively deep clefts which provide access into the inner regions of the ovarian ball (Meyer, 1933; Crompton and Whitfield, 1974; Atkinson and Byram, 1976). In the eoacanthocephalans *Aca. tilapiae* and *Pal. golvani*, Marchand and Mattei (1976d) showed that the spermatozoa became attached to the ovarian balls by their free flagella. They also described a centriolar derivative (La baguette) in *S. socialis* which was believed to function rather like an acrosome in that it facilitated the entry of the spermatozoa into the ovarian ball (Marchand and Mattei, 1977a). Over 100 sections through spermatozoa were counted by them in an ultra-thin section through an ovarian ball of *Aca. tilapiae* (Marchand and Mattei, 1976d). The sections of spermatozoa within the ovarian ball were usually cut through the free flagellum and revealed that the spermatozoon was generally separated from the surrounding ovarian tissue by some form of extracellular space. Spermatozoan sections were observed in the supporting syncytium, the oogonial syncytium and the oocytes; electron micrographs showed the spermatozoon to be isolated from the oocyte, which was apparently surrounding it, by a thin collar or sleeve of supporting syncytial tissue (Marchand and Mattei, 1976d). The significance of these complex spatial relationships between the spermatozoa and the components of the ovarian ball is not understood, but they may represent a stage in the incorporation of spermatozoa into the mature oocytes of *Aca. tilapiae* and *Pal. golvani*. It may be unwise to link the function of the centriolar derivative in *S. socialis* with that of a true acrosome which is considered to be involved in the interaction of a spermatozoan and oocyte during fertilization (Austin, 1965); the observations of Marchand and Mattei (1976d, 1977a) are mostly concerned with interactions between spermatozoa and the supporting syncytium which is not a part of the germ line in function.

Crompton and Whitfield (1974) published an electron micrograph of a section through an oocyte of *P. minutus* in which the axoneme of a spermatozoan flagellum is apparent. This structure must be part of a spermatozoon because evidence for the formation of the vitelline membrane can be seen (see below) in the micrograph and fertilization, as defined above, must have been occurring when the material was fixed. This single observation of a naked spermatozoan axoneme in direct contact with the oocytic cytoplasm might indicate that the spermatozoa of *P. minutus* may interact with the oocytes by a fusion of plasma membranes followed by an intermingling of nuclear and cytoplasmic contents. Speculation about this aspect of acanthocephalan fertilization should be withheld until further studies, preferably *in vitro* where a time course could be determined, have been carried out. Dunagan (1962) appears to have made some useful progress towards developing an *in vitro* aseptic technique for studying fertilization in *Neoechinorhynchus* spp. from turtles. Copulation occurred and motile spermatozoa, seen to be attached to the surface of ovarian balls and eggs, were produced; there is, however, the possibility that the female worms had already been inseminated before their removal from the turtles.

Several lines of evidence indicate that activation of the egg or mature oocyte occurs after the entry of the spermatozoon. The cytoplasm of the syncytium and its derivatives in *P. minutus* and *M. dubius* contains electron-dense, membrane-bounded inclusions which appear to have become most complex and numerous in the cytoplasm of the mature oocytes (Crompton and Whitfield, 1974). The inclusions have been observed to be distributed either evenly throughout the cytoplasm of the mature oocytes or more densely in the peripheral cytoplasm. At about the same time as the distribution of the inclusions is altered, the oocyte changes from a spherical to an oval shape, an expansion of the space occurs between the oocyte's surface and the supporting syncytium, and the fertilization membrane, which becomes part of the egg shell, is formed (Whitfield, 1973; Crompton and Whitfield, 1974; Atkinson and Byram, 1976). These events are reminiscent of accounts of the cortical reaction during fertilization in other groups of invertebrates (see Austin, 1965) and they probably occur also in *Pom. laevis* (Stranack, 1972), *Prosthenorchis* sp. (Guraya, 1969) and *C. corvi* (Parshad and Guraya, 1977b). Histochemical tests indicate staining differences in the cytoplasm of the mature oocytes before and after this activation phase of fertilization (Guraya, 1969; Parshad and Guraya, 1977b).

A few scattered observations of relevance to the fusion of the male and female pronuclei are available. For example, Robinson (1965) concluded that the nuclear material in the threadlike spermatozoon of *M. dubius* condensed on entry into the mature oocyte whose nucleus was then stimulated to divide meiotically. Presumably one of these haploid female nuclei would subsequently fuse with the male nucleus and the polar bodies would be formed. Crompton and Whitfield (1974) also considered that the nuclei of the mature oocytes of *M. dubius* and *P. minutus* were in the premeiotic condition before fertilization, but an observation of Atkinson and Byram (1976) indicated that the nuclei of some of the oocytes in 12-day-old *M. dubius* were undergoing meiotic prophase. Evidence for the occurrence of fertilization in 12-day-old *M. dubius* has not been obtained (Crompton, 1974; Atkinson and Byram, 1976). The diagrams of Nicholas and Hynes (1963) and Schmidt (1973) for *P. minutus* and *Med. grandis* respectively depict two polar bodies in association with the zygote whereas Robinson (1965) photographed three in an equivalent stage of *M. dubius*. It is not unusual for some metazoan groups to produce two polar bodies and others to form three (Austin, 1965); the results of comparative studies on other species of Acanthocephala would be of interest.

Various changes occur in the females of *M. dubius* following the completion of fertilization. In addition to embryonic development, which takes place in the body cavity (see below), the somatic tissues of female worms grow bigger than those of uninseminated females (Crompton, 1972, 1974). A comparison of the mean amount of nitrogen in female *M. dubius* aged from 5 to 15 weeks showed that there was always more nitrogen in the bodies of inseminated females than in entire uninseminated females of the same age.

VIII. Development of the Egg (Shelled Acanthor)

Many workers use the term egg for brevity and convenience to describe the acanthor larva enclosed by an egg shell. In fact, the shell is better considered as a series of envelopes (Wright, 1971) and it is suggested that these should be identified by numbers rather than names in an attempt to minimize confusion and avoid error (Crompton, 1970; Whitfield, 1973). The outer envelope should be number 1, followed by number 2 and so on. After the completion of fertilization, the zygote becomes detached from the surface of the ovarian ball and embryonic development continues in the body cavity (Nicholas, 1967). The egg envelopes are also formed during this time without participation of any special moulding apparatus of the type found in platyhelminths (Van Cleave, 1953). Most accounts of studies of acanthocephalan life cycles in the laboratory indicate that intermediate hosts may become infected by eating eggs taken by the observer directly from the body cavity of a female worm. Many of these eggs will not be fully developed (Fig. 8A), particularly if taken from a relatively young female, but this general observation shows that the fully-developed egg leaves the parent without any need for a period of development in the environment of the host. The fully developed eggs of several species are shown in Fig. 19 and further information about shape, size and number of envelopes is given in Table 10. Although not exactly within the scope of this review, it is worth noting that acanthocephalan eggs are often adapted to withstand a great variety of conditions in the host's environment without losing their infectivity (see Crompton, 1970). The egg of *Mac. hirudinaceus* appears to be particularly hardy (Kates, 1942), presumably because of the properties of the 4 envelopes (Table 10) although these are not likely to protect the acanthor from the effects of freezing (Kates, 1942).

A. EMBRYONIC DEVELOPMENT

A few observations about the time taken for embryonic development to occur in four species of Acanthocephala are given in Table 9 and studies of the embryology of the following species have been made: Palaeacanthocephala, *A. ranae* (Hamann, 1891), *E. gadi* (Hamann, 1891) and *P. minutus* (Nicholas and Hynes, 1963); Archiacanthocephala, *Mac. hirudinaceus* (Meyer, 1928, 1936, 1937, 1938a, b), *M. dubius* (Nicholas, 1967) and *Med. grandis* (Schmidt, 1973); Eoacanthocephala, *N. rutili* (Meyer, 1931). Schmidt (1973) has concluded that the developmental pattern is similar in all these species with the cleavage being of a distorted spiral type. Cleavage may begin while the zygote is associated with the ovarian ball, but it is more readily

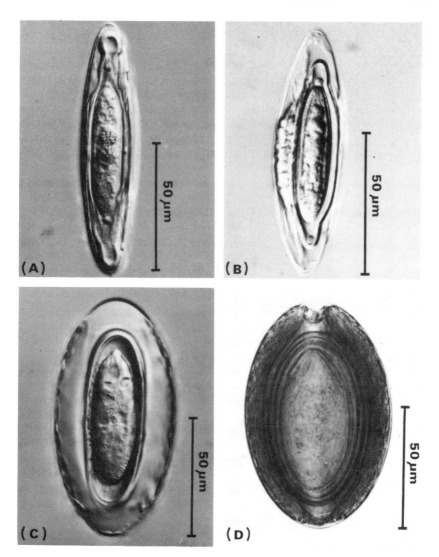

FIG. 19. Photomicrographs of acanthocephalan eggs (shelled acanthors) taken from the body cavities of mature female worms. (A) *Polymorphus minutus* (Pa., Nomarski). (B) *Acanthocephalus ranae* (Pa., Nomarski). (C) *Moniliformis dubius* (Ar., Nomarski). (D) *Macracanthorhynchus hirudinaceus* (Ar., bright field). Ar., Archiacanthocephala; Pa., Palaeacanthocephala.

observed in stages that are free in the body cavity (Nicholas and Hynes, 1963). The first two cleavages are slightly unequal to produce four macromeres which divide asymmetrically to produce macromeres and micromeres. After

TABLE 10

Observations on the eggs (shelled acanthors) of Acanthocephala

Parasite	Dimensions[a,b] (μm)		Number of envelopes	References
ARCHIACANTHOCEPHALA				
Macracanthorhynchus hirudinaceus (m)[c]	l.	80–100	4	Meyer (1933); Kates (1943)
	w.	46–65		
Mediorhynchus centurorum (b)	l.	55	4(3)[d]	Nickol (1969, 1977)
	w.	36		
Med. grandis (b)	l.	58–64	4	Moore (1962); Schmidt (1973)
	w.	35–38		
Moniliformis clarki (b)	l.	60–75	4	Crook and Grundmann (1964)
	w.	35–40		
M. dubius (m)	l.	112–120	4	Moore (1946); Edmonds (1966)*; Wright (1971)*
	w.	56–60		
PALAEACANTHOCEPHALA				
Acanthocephalus jacksoni (f)	l.	65–103	4(3)[d]	Bullock (1962); West (1963, 1964)*; Oetinger and Nickol(1974)
	w.	14–19		
Echinorhynchus lageniformis (f)	l.	90	4	Olson and Pratt (1971)
	w.	20		
E. truttae (f)	l.	110–140	3(4)[e]	Awachie (1966)
	w.	23–27		
Fessisentis fessus (a)	l.	111–124	4	Nickol (1972); Dunagan and Miller (173); Buckner and Nickol (1979)
	w.	14–19		
F. vancleavi (a)	l.	68–83	4	Buckner and Nickol (1978)
	w.	13–15		
Leptorhynchoides thecatus (f)	l.	85	4(3)[d]	DeGiusti (1949); Uznanski and Nickol (1976)
	w.	21		

TABLE 10 (*continued*)

Polymorphus minutus (b)	l. 96–109 w. 18–20	3(4)[e]	Monné and Hönig (1954)*; Whitfield (1973)*
Prosthorhynchus formosus (b)	l. 75 w. 30	3	Schmidt and Olsen (1964)
EOACANTHOCEPHALA			
Acanthosentis digospinus (f)	n.g.	3	Anantaraman and Ravindranath (1976)*
Neoechinorhynchus cristatus (f)	l. 56 w. 27	4	Uglem (1972)
N. emydis (r)	l. 25 w. 18–22	3	Hopp (1954)
N. rutili (f)	l. 27 w. 17	3	Merritt and Pratt (1964)
N. saginatus (f)	w. 44–46 w. 16–20	4	Uglem and Larson (1969)
Octospinifer macilentis (f)	l. 51–60 w. 23–28	3	Harms (1965)
Pallisentis nagpurensis (f)	l. 92 w. 48	3	George and Nadakal (1973)

[a] Measurements were made on eggs obtained from various locations including host faeces, the body cavities of both living and fixed worms and the efferent duct system. l, length; w, width.

[b] Egg sizes vary according to location (see Nickol, 1972; Buckner and Nickol, 1978, 1979).

[c] Definitive host: a, amphibian; b, bird; f, fish; m, mammal; r, reptile.

[d] Outer envelope may be lost once the egg has been passed out of the host.

[e] Middle envelope may be further resolved into two components.

* References give details of the structure and composition of the egg envelopes.

n.g. Measurements not given.

relatively few divisions, cell membranes can no longer be observed by means of conventional light microscopy and the embryo appears to be syncytial. In general, the polar bodies are located at the anterior end of the developing embryo (Nicholas and Hynes, 1963; Schmidt, 1973).

The fully developed acanthor contains a central mass of nuclei (Fig. 2A) which appears to originate from the micromeres and continues to divide. Nicholas (1967), following the views of Meyer, considers that the formation of the central nuclear mass is the equivalent of gastrulation in other animals. These central nuclei give rise to the rudiments of the ganglion, ligament, body muscles and gonads. Not all the embryonic nuclei become assembled in the central cytoplasm of the acanthor. In some species, nuclei derived from the micromeres, which were formed during subsequent divisions of the four macromeres, become the nuclei of the body wall. Schmidt (1973) has estimated that there are about 180–200 nuclei in the central nuclear mass and about 30–40 surrounding nuclei in the embryonic *Med. grandis*.

B. EGG ENVELOPES

The acanthors of most acanthocephalan species are probably surrounded at some stage of development by four envelopes (Fig. 19; Table 10), despite the fact that only three are often observed. Four envelopes are usually seen when eggs are removed directly from the body cavity of the female and examined with the light microscope. The thin outer layer (envelope 1) may be lost sometime after the egg leaves the body cavity of the female; in some palaeacanthocephalan eggs, envelope 1 has a tendency to disintegrate into large fibrils which may anchor the eggs to aquatic vegetation or cause them to stick together (Oetinger and Nickol, 1974; Uznanski and Nickol, 1976). Most information is available about the egg envelopes of *M. dubius* (Wright, 1971) and *P. minutus* (Whitfield, 1973) whose papers should be consulted.

The outer envelope of the egg of *Pom. laevis*, *M. dubius* and *P. minutus* is fertilization membrane which is produced during fertilization by a type of cortical reaction in the mature oocyte (Section VII; Stranack, 1972; Whitfield, 1973; Crompton and Whitfield, 1974; Atkinson and Byram, 1976). Information about the formation of envelopes 2, 3 and 4 is inadequate, but it seems likely that they are produced by the developing embryo with the adult worm contributing the precursor materials.

This assumption means that the embryo may be solely involved in synthesizing chitin (Von Brand, 1940; Monné and Hönig, 1954; Edmonds, 1966; Wright, 1971) and keratin (Monné and Hönig, 1954; Whitfield, 1973) as well as other components which have tentatively been identified as tunicin, cellulose, elastin-like protein, fibrin, polyphenols and acid mucopolysaccharide (see Whitfield, 1973; Anantaraman and Ravindranath, 1976).

IX. EGG RELEASE AND PATENCY

There is a dearth of information about egg production by Acanthocephala (Table 10). Most observations have been made on *M. dubius* and *P. minutus* and it is clear from the references cited in Table 10 that several factors may affect both the beginning of egg release and its duration. Nicholas and Hynes (1958) concluded that the release of eggs by female *P. minutus* might not be a continuous process. They described how they failed to find eggs of *P. minutus* in the faeces of three ducks which were subsequently found to contain many gravid female and mature worms. Their unsuccessful search was made from day 21 to day 25 after the oral infection of the ducts. In a similar study, Crompton and Whitfield (1968) first observed eggs of *P. minutus* in duck faeces 22 days after the infection of the ducks. This interesting difference in experimental results warrants further study because it may indicate that a rhythmical pattern of egg release exists in *P. minutus*. Nicholas and Hynes (1958) had no difficulty in finding eggs of *P. minutus* which they had already added to duck faeces. The observations of Awachie (1966) on egg release by *E. truttae* raise another intriguing question. Apparently, the patent period of *E. truttae* lasts for only about a third of the length of the prepatent period (Table 10). Finally, studies need to be carried out on the release of eggs by eoacanthocephalans. It should be remembered, however, that any observations made on egg release by acanthocephalans in controlled laboratory studies will not necessarily facilitate our understanding of egg release by worms in natural infections, in which the recruitment and loss of worms of both sexes to and from an established population will probably occur.

X. SUGGESTIONS FOR FURTHER WORK

We decided that to make any firm conclusions about acanthocephalan reproduction would be premature and probably misleading for any readers who are not directly involved in research on the group. Instead, we have compiled a list of some new topics for research and some old problems for re-investigation. The list is not intended to be comprehensive; it reflects our present enthusiasms and interests which have been stimulated by the opportunity to survey some of the literature during the preparation of the review. The list also cites some of the published work which could serve as a starting point for further study.

1. Is there any reproductive significance in the morphological somatic differences observed between male and female acanthocephalans of the same species?

2. Are the size differences between male and female worms genuine or do they result from nutritional and other environmental factors?

3. What is the significance of the chemical and metabolic differences observed between males and females of the same species? (Graff, 1964; Rothman and Fisher, 1964).

4. Is the XO mechanism of sex determination more common than the XY mechanism? (Robinson, 1964, 1965).

5. What are the karyotypes of species of Acanthocephala in addition to those mentioned in Table 3?

6. What is the karyotype of *P. minutus* (Table 3)?

7. Do female acanthocephalans usually live longer than males of the same species?

8. What factors other than ambient temperature affect acanthocephalan development in the intermediate host?

9. Do any of the nuclei of the central nuclear mass (Fig. 2A) possess the capacity for giving rise to the germ-line primordia?

10. What is the role of the embryonic cytoplasm during the formation of the germ-line primordia?

11. What is the role of the ligament in the development of the reproductive system?

12. At what stage during development are the first ovarian balls formed from the ovarian rudiment? (Yamaguti and Miyata, 1942; Van Cleave, 1953; Schmidt and Olsen, 1964; Atkinson and Byram, 1976). The "first ovarian balls" should be considered as immature; the functional ovarian ball is an ovary (Atkinson and Byram, 1976).

13. How common is monorchidism in Acanthocephala? (Bullock, 1962; Buckner and Nickol, 1978).

14. What is the composition of the secretion(s) of the cement gland(s)?

15. Does each gland produce the same secretion?

16. What is the function of the other gland-like structures in the male reproductive tract? (Dunagan and Miller, 1973).

17. What is the function of Saefftigen's pouch?

18. What is the average number of ovaries (=mature ovarian balls) in female Acanthocephala of a given species during the course of an infection?

19. What factors affect the numbers and sizes of ovaries during the course of an infection?

20. Does the number of ovaries increase as the result of the division of the existing population of ovaries?

21. If so, how does the process of division occur?

22. Is each ovary capable of division?

23. How common are the multi-unit ovaries of the type found in *C. corvi*? (Parshad and Guraya, 1977b).

24. What additional evidence can be obtained to test the view that the uterine bell functions as an egg-sorting device? (Whitfield, 1968, 1970).

25. How are mature eggs (=shelled acanthors) distinguished from developing eggs in the uterine bell?

26. How does the functional testis develop?

27. What are the origins of the germ-line and supporting cells in the testis? (Whitfield, 1969).

28. Do seasonal changes occur in acanthocephalan testes and if so can these be related to the host's reproductive cycle? (Parshad and Guraya, 1979).

29. How soon after the infection of the definitive host are mature spermatozoa formed?

30. What factors affect spermatogenesis?

31. Is there an acrosome or its equivalent in acanthocephalan spermatozoa? (Whitfield, 1971a; Marchand and Mattei, 1977a).

32. What are the origins of the supporting and oogonial syncytia in the ovaries? (Atkinson and Byram, 1976).

33. How widespread is oocyte atresia and how frequently and under what conditions does it occur in a given species? (Parshad and Guraya, 1978).

34. When does copulation first occur during the course of infection in different species of Acanthocephala?

35. How do male and female worms come into contact?

36. What is the significance of the copulation cap? (Van Cleave, 1949a; Abele and Gilchrist, 1977).

37. How are spermatozoa transferred from the male to the female? Perhaps some form of semen is involved and if so where is it produced and what is its composition in addition to spermatozoa?

38. Is there any direct evidence for the existence of spermatophores in the Acanthocephala? (Atkinson and Byram, 1976).

39. For how long can spermatozoa live inside the body cavity of the inseminated female?

40. Do the females of any species possess organs or structures for the storage of spermatozoa?

41. What is happening to the spermatozoa that are observed at the surfaces of the ovaries?

42. How is fertilization achieved?

43. Does fertilization always occur while the oocytes are contained within the ovary?

44. How do spermatozoa pass through the supporting syncytium at the ovarian surface?

45. How do spermatozoa enter the mature oocytes?

46. How much of the spermatozoon enters the oocyte?

47. How do the male and female nuclear materials fuse?

48. How widespread is the 'cortical reaction' of the type described for *M. dubius* and *P. minutus* and does it function as a block to polyspermy?

49. How many polar bodies are formed in different species of Acanthocephala? (Robinson, 1965; Schmidt, 1973).

50. How is the zygote or embryo released from the ovary into the body cavity?

51. When is the embryo released from the ovary?

52. How long does embryonic development take in different species of Acanthocephala? (Table 9).

53. What factors affect the duration of embryonic development?

54. What is the fate of the cleavage products during embryonic development?

55. How are the different egg envelopes (Table 10) formed during development?

56. How soon after the infection of a definitive host by a given species does egg release begin?

57. What is the duration of egg release by a given species?

58. Is there any evidence for a rhythmical or seasonal pattern of egg release?

59. What factors affect the production of eggs?

60. How many eggs are produced on average by a given species of Acanthocephala?

Attention should also continue to be given to studies of the functional morphology of the reproductive tract, to observations on the course of infection and to cellular and biochemical aspects of reproduction. It is most important that comparative investigations should be carried out on as many representative species as possible from the three orders of the phylum.

ACKNOWLEDGEMENTS

We are most grateful to Dr. W. B. Amos, Mr. J. Brattey, Mr. P. Curtis, Dr. J. R. Georgi, Dr. B. Marchand and Dr. P. J. Whitfield for help and advice during the preparation of this review. We also thank Professor S. S. Guraya and Dr. B. A. Newton for providing facilities and support for our work, and the University of Cambridge, the Commonwealth Scholarship and Fellowship Scheme and the Royal Society–Indian National Science Academy exchange programme for financial support. Full acknowledgement to the sources of tables and illustrations which have been used in the review has been given in the text. Finally, it is a pleasure to thank Mrs. Rina Clark and Miss Paula Johnson for their careful preparation of the typescript.

REFERENCES

Abele, L. G. and Gilchrist, S. (1977). Homosexual rape and sexual selection in acanthocephalan worms. *Science* **197,** 81–83.

Ali, S. (1977). "Field Guide to the Birds of the Eastern Himalayas." Oxford University Press.

Amin, O. M. (1975a). *Acanthocephalus parksidei* sp.n. (Acanthocephala: Echinorhynchidae), from Wisconsin fishes. *Journal of Parasitology* **61,** 301–306.

Amin, O. M. (1975b). Variability in *Acanthocephalus parksidei,* Amin, 1974 (Acanthocephala: Echinorhynchidae). *Journal of Parasitology* **61,** 307–317.

Anantaraman, S. and Ravindranath, M. H. (1976). Histochemical characteristics of the egg envelopes of *Acanthosentis* sp. (Acanthocephala). *Zeitschrift für Parasitenkunde* **48,** 227–238.

Anantaraman, S. and Subramoniam, T. (1975). Oogenesis in *Acanthosentis oligospinus* n.sp., an aganthocephalan parasite of the fish, *Macrones gulio. Proceedings of the Indian Academy of Sciences* **82B,** 139–145.

Asaolu, S. O. (1976). Ovarian ball development in *Moniliformis* (Acanthocephala). *Parasitology* **73,** xxvii.

Asaolu, S. O. (1977). Studies on the Reproductive Biology of *Moniliformis* (Acanthocephala). PhD. Dissertation, University of Cambridge.

Asaolu, S. O. (1980). Morphology study of the reproductive system of female *Moniliformis* (Acanthocephala). *Parasitology* **81,** 433–446.

Atkinson, K. H. and Byram, J. E. (1976). The structure of the ovarian ball and oogenesis in *Moniliformis dubius* (Acanthocephala). *Journal of Morphology* **148,** 391–426.

Austin, C. R. (1965). "Fertilization". Prentice-Hall, Englewood Cliffs, New Jersey.

Awachie, J. B. E. (1966). The development and life history of *Echinorhynchus truttae* Schrank, 1788 (Acanthocephala). *Journal of Helminthology* **40**, 11–32.

Baer, J. C. (1961). Embranchement des Acanthocéphales. *In* "Traité de Zoologie" vol. IV (P. P. Grassé, ed.), Masson et Cie, Paris.

Beames, C. G. and Fisher, F. M. (1964). A study on the neutral lipids and phospholipids of the Acanthocephala *Macracanthorhynchus hirudinaceus* and *Moniliformis dubius*. *Comparative Biochemistry and Physiology* **13**, 401–412.

Bloom, W. and Fawcett, D. W. (1968). "A Textbook of Histology". W. B. Saunders Co., Philadelphia and London.

Bone, L. W. (1974a). The chromosones of *Leptorhynchoides thecatus* (Acanthocephala). *Journal of Parasitology* **60**, 818.

Bone, L. W. (1974b). The chromosomes of *Neoechinorhynchus cylindratus* (Acanthocephala). *Journal of Parasitology* **60**, 731–732.

Brand, T. von (1939). Chemical and morphological observations upon the composition of *Macracanthorhynchus hirudinaceus* (Acanthocephala). *Journal of Parasitology* **25**, 329–342.

Brand, T. von (1940). Further observations on the composition of Acanthocephala. *Journal of Parasitology* **26**, 301–307.

Brattey, J. (1980). Preliminary observations on larval *Acanthocephalus lucii* (Müller, 1776) (Acanthocephala: Echinorhynchidae) in the isopod *Asellus aquaticus* (L.). *Parasitology* **81**, xlix.

Buckner, S. C. and Nickol, B. B. (1975). Morphological variation of *Moniliformis moniliformis* (Bremser, 1811), Travassos, 1915 and *Moniliformis clarki* (Ward 1917), Chandler, 1921. *Journal of Parasitology* **61**, 996–998.

Buckner, R. L. and Nickol, B. B. (1978). Redescription of *Fessisentis vancleavei*, (Hughes and Moore 1943) Nickol 1972. (Acanthocephala: Fessisentidae). *Journal of Parasitology* **64**, 635–637.

Buckner, R. L. and Nickol, B. B. (1979). Geographic and host-related variation among species of *Fessisentis* (Acanthocephala) and confirmation of the *Fessisentis fessus* life cycle. *Journal of Parasitology* **65**, 161–166.

Bullock, W. L. (1958). Histochemical studies on the Acanthocephala. III. Comparative histochemistry of alkaline glycerophosphatase. *Experimental Parasitology* **7**, 51–68.

Bullock, W. L. (1962). A new species of *Acanthocephalus* from New England fishes, with observations on variability. *Journal of Parasitology* **48**, 442–451.

Bullock, W. L. (1969). Morphological features as tools and as pitfalls in Acanthocephalan systematics. *In* "Problems in Systematics of Parasites" (G. D. Schmidt, ed.), University Park Press, Baltimore, Maryland and Manchester.

Burlingame, P. L. and Chandler, A. C. (1941). Host-parasite relations of *Moniliformis dubius* (Acanthocephala) in albino rats, and the environmental nature of resistance to single and superimposed infections with this parasite. *American Journal of Hygiene* **33**, 1–21.

Cable, R. M. and Dill, W. T. (1967). The morphology and life history of *Paulisentis fractus* Van Cleave and Bangham, 1949. (Acanthocephala: Neoechinorhynchidae). *Journal of Parasitology* **53**, 810–817.

Calow, P. (1978). "Life Cycles". Chapman and Hall, London.

Cohen, J. (1977). "Reproduction". Butterworths, London.

Crompton, D. W. T. (1970). "An Ecological Approach to Acanthocephalan Physiology". Cambridge University Press.

Crompton, D. W. T. (1972a). The growth of *Moniliformis dubius* (Acanthocephala) in the intestine of male rats. *Journal of Experimental Biology* **56**, 19–29.

Crompton, D. W. T. (1972b). Monorchic *Moniliformis dubius* (Acanthocephala). *International Journal for Parasitology* **2**, 483.

Crompton, D. W. T. (1974). Experiments on insemination in *Moniliformis dubius* (Acanthocephala). *Parasitology* **68**, 229–238.

Crompton, D. W. T. (1975). Relationships between Acanthocephala and their hosts. *In* "Symbiosis" (D. H. Jennings and D. L. Lee, eds.), Symposium of the Society for Experimental Biology 29.

Crompton, D. W. T. and Walters, D. E. (1972). An analysis of the course of infection of *Moniliformis dubius* (Acanthocephala) in rats. *Parasitology* **64**, 517–523.

Crompton, D. W. T. and Whitfield, P. J. (1968). The course of infection and egg production of *Polymorphus minutus* (Acanthocephala) in domestic ducks. *Parasitology* **58**, 231–246.

Crompton, D. W. T. and Whitfield, P. J. (1974). Observations on the functional organization of the ovarian balls of *Moniliformis* and *Polymorphus* (Acanthocephala). *Parasitology* **69**, 429–443.

Crompton, D. W. T., Arnold, S. and Barnard, D. (1972). The patent period and production of eggs of *Moniliformis dubius* (Acanthocephala) in the small intestine of male rats. *International Journal for Parasitology* **2**, 319–326.

Crompton, D. W. T., Arnold, S. and Walters, D. E. (1976). The number and size of ovarian balls of *Moniliformis* (Acanthocephala) from laboratory rats. *Parasitology* **73**, 65–72.

Crook, J. R. and Grundmann, A. W. (1964). The life history and larval development of *Moniliformis clarki* (Ward, 1917). *Journal of Parasitology* **50**, 689–693.

Davenport, R. (1979). "An Outline of Animal Development". Addison-Wesley Publishing Company, Reading, Massachusetts and London.

DeGiusti, D. L. (1949). The life cycle of *Leptorhynchoides thecatus* (Linton), an acanthocephalan of fish. *Journal of Parasitology* **35**, 437–460.

Dunagan, T. T. (1962). Studies on *in vitro* survival of Acanthocephala. *Proceedings of the Helminthological Society of Washington* **29**, 131–135.

Dunagan, T. T. and Miller, D. M. (1973). Some morphological and functional observations on *Fessisentis fessus* Van Cleave (Acanthocephala) from the Dwarf Salamander, *Siren intermedia* Le Conte. *Proceedings of the Helminthological Society of Washington* **40**, 209–216.

Dunagan, T. T. and Miller, D. M. (1977). A new ganglion in male *Moniliformis moniliformis* (Acanthocephala). *Journal of Morphology* **152**, 171–176.

Dunagan, T. T. and Miller, D. M. (1978a). Muscles of the reproductive system of male *Moniliformis moniliformis* (Acanthocephala). *Proceedings of the Helminthological Society of Washington* **45**, 69–76.

Dunagan, T. T. and Miller, D. M. (1978b). Anatomy of the genital ganglion of the male Acanthocephalan, *Moniliformis moniliformis*. *Journal of Parasitology* **64**, 431–435.

Dunagan, T. T. and Miller, D. M. (1979). Genital ganglion and associated nerves in male *Macracanthorhynchus hirudinaceus* (Acanthocephala). *Proceedings of the Helminthological Society of Washington* **46**, 106–114.

Edmonds, S. J. (1966). Hatching of the eggs of *Moniliformis dubius*. *Experimental Parasitology* **19**, 216–226.

Fisher, F. M. (1960). On Acanthocephala of turtles, with the description of *Neoechinorhynchus emyditoides* n.sp. *Journal of Parasitology* **46**, 257–266.

Fotedar, D. N. (1968). New species of *Neoechinorhynchus*, Hamann, 1892 from *Oreinus sinuatus*, freshwater fish in Kashmir. *Kashmir Science* **5**, 147–152.

Furtado, J. I. (1963). On *Acanthogyrus partispinus* nov. sp. (Quadrigyridae, Acanthocephala) from a Malayan cyprinid *Hampala macrolepidota* Van Hasselt. *Zeitschrift für Parasitenkunde* **23**, 219–225.

George, P. V. and Nadakal, A. M. (1973). Studies on the life cycle of *Pallisentis nagpurensis* Bhalerao, 1931 (Pallisentidae: Acanthocephala) parasitic in the fish *Ophiocephalus striatus* (Bloch). *Hydrobiologia* **42**, 31–43.

Golvan, Y. J. (1956). Acanthocéphales d'oiseaux. Note additionelle. *Pseudoporrorchis rotundatus* (O. V. Linstow, 1897) (Palaeacanthocephala-Polymorphidae), parasite d'un Cucullidae, *Centropus madagascariensis* (Briss). *Bulletin de la Société zoologique de France* **81**, 339–344.

Graff, D. J. (1964). Metabolism of C^{14}-Glucose by *Moniliformis dubius* (Acanthocephala). *Journal of Parasitology* **50**, 230–234.

Graff, D. J. and Allen, K. (1963). Glycogen content in *Moniliformis dubius* (Acanthocephala). *Journal of Parasitology* **49**, 204–208.

Guraya, S. S. (1969). Histochemical observations on the developing acanthocephalan oocyte. *Acta Embryologiae Experimentalis* **1**, 147–155.

Guraya, S. S. (1971). Morphological and histochemical observations on the acanthocephalan spermatogenesis. *Acta Morphologica Neerlando-Scandinavica* **9**, 75–83.

Haley, A. J. and Bullock, W. L. (1952). Comparative histochemical studies on the cement glands of certain Acanthocephala. *Journal of Parasitology* **38**, 25–26.

Hamann, O. (1891). Monographie der Acanthocephalen (Echinorhynchen) *Jenaische Zeitschrift für Naturwissenschaft* **25**, 113–232.

Harada, I. (1931). Das Nervensystem von *Bolbosoma turbinella* (Dies). *Japanese Journal of Zoology* **3**, 161–199.

Harms, C. E. (1965). The life cycle and larval development of *Octospinifer macilentis* (Acanthocephala: Neoechinorhynchidae). *Journal of Parasitology* **51**, 286–293.

Holmes, J. C. (1961). Effects of concurrent infections on *Hymenolepis diminuta* (Cestoda) and *Moniliformis dubius* (Acanthocephala). I. General effects and comparison with crowding. *Journal of Parasitology* **47**, 209–216.

Holloway, H. L. and Nickol, B. B. (1970). Morphology of the trunk of *Corynosoma hamanni* (Acanthocephala: Polymorphidae). *Journal of Morphology* **130**, 151–162.

Hopp, W. B. (1954). Studies on the morphology and life cycle of *Neoechinorhynchus emydis* (Leidy), an acanthocephalan parasite of the map turtle, *Graptemys geographica* (Le Sueur). *Journal of Parasitology* **40**, 284–299.

Hyman, L. H. (1951). "The Invertebrata", vol. III. McGraw-Hill, New York, Toronto and London.

Hynes, H. B. N. and Nicholas, W. L. (1957). The development of *Polymorphus minutus* (Goeze, 1782) (Acanthocephala) in the intermediate host. *Annals of Tropical Medicine and Parasitology* **51**, 380–391.

Jones, A. W. and Ward, H. L. (1950). The chromosomes of *Macracanthorhynchus hirudinaceus* (Pallas). *Journal of Parasitology* **36**, 86.

Johnston, T. H. and Deland, E. W. (1929). Australian Acanthocephala, No. 2. *Transactions of the Royal Society of South Australia* **53**, 155–166.

Kaiser, J. E. (1893). Die Acanthocephalen und ihre Entwicklung. *Bibliotheca Zoologica* **11**, Heft 7.

Kates, K. C. (1942). Viability of the eggs of the swine thorn-headed worm (*Macracanthorhynchus hirudinaceus*). *Journal of Agricultural Research* **64**, 93–100.

Kates, K. C. (1943). Development of the swine thorn-headed worm *Macracanthorhynchus hirudinaceus* in its intermediate host. *American Journal of Veterinary Research* **4**, 173–181.

Kates, K. C. (1944). Some observations on experimental infections of pigs with the thorn-headed worm, *Macracanthorhynchus hirudinaceus*. *American Journal of Veterinary Research* **5**, 166–172.

Kennedy, C. R. (1972). The effects of temperature and other factors upon the establishment and survival of *Pomphorhynchus laevis* (Acanthocephala) in goldfish, *Carassius auratus*. *Parasitology* **65**, 283–294.

Kennedy, C. R. (1975). "Ecological Animal Parasitology". Blackwell Scientific Publications, Oxford, London, Edinburgh and Melbourne.

Khatkevich, L. M. (1975a). The histological structure of the uterine bell, uterus and vagina of the acanthocephalan *Echinorhynchus gadi* Mueller, 1776 (In Russian). Original article not seen. *Helminthological Abstracts* **48**, 3352.

Khatkevich, L. M. (1975b). The histological structure of the efferent duct in female *Neoechinorhynchus rutili* (Mueller, 1780) Hamann, 1892 (Acanthocephala) (In Russian). Original article not seen. *Helminthological Abstracts* **47**, 503.

Kilian, R. (1932). Zur Morphologie und Systematik der Giganthorhynchidae (Acanthocephala). *Zeitschrift für wissenschaftliche Zoologie* **141**, 246–345.

King, D. and Robinson, E. S. (1967). Aspects of the development of *Moniliformis dubius*. *Journal of Parasitology* **53**, 142–149.

Kobayashi, M. (1959). Studies on the Acanthocephala (3). Studies on the anomaly of testis of *Centrorhynchus elongatus* Yamaguti. *Japanese Journal of Parasitology* **8**, 423.

Lackie, J. M. (1972). The course of infection and growth of *Moniliformis dubius* (Acanthocephala) in the intermediate host *Periplaneta americana*. *Parasitology* **64**, 95–106.

Lewis, K. R. and John, B. (1968). The chromosomal basis of sex determination. *International Review of Cytology* **23**, 277–379.

Marchand, B. and Mattei, X. (1976a). La spermatogenèse des Acanthocéphales. 1. L'appareil centriolaire et flagellaire au cours de la spermiogenese d'*Illiosentis furcatus* var *africana* Golvan, 1956 (Paleacanthocephala, Rhadinorhynchidae). *Journal of Ultrastructure Research* **54**, 347–358.

Marchand, B. and Mattei, X. (1976b). La spermatogenèse des Acanthocéphales. 2. Variation du nombre des fibres centrales dans la flagelle spermatique d'*Acanthosentis tilapiae* Baylis, 1947 (Eoacanthocephala, Quadrigyridae). *Journal of Ultrastructure Research* **55**, 391–399.

Marchand, B. and Mattei, X. (1976c). Ultrastructure du spermatozoide de *Centrorhynchus milvus* Ward, 1956 (Paleacanthocephala, Polymorphidae). *Comptes Rendus des Séances de la Société Biologie* **170**, 237–240.

Marchand, B. and Mattei, X. (1976d). Présence de flagelles spermatiques dans les sphères ovariennes des eoacanthocéphales. *Journal of Ultrastructure Research* **56**, 331–338.

Marchand, B. and Mattei, X. (1977a). La spermatogenèse des Acanthocéphales. 3. Formation du dérivé centriolaire au cours de la spermiogenèse de *Serrasentis socialis* Van Cleave, 1924 (Paleacenthocephala, Gorgorhynchidae). *Journal of Ultrastructure Research* **59**, 263–271.

Marchand, B. and Mattei, X. (1977b). Un type nouveau de structure flagellaire. Type 9+n. *Journal of Cell Biology* **72**, 707–713.

Marchand, B. and Mattei, X. (1978a). La spermatogenèse des Acanthocéphales. 4. Le dérivé nucléocytoplasmique. *Biologie Cellulaire* **31**, 79–90.

Marchand, B. and Mattei, X. (1978b). La spermatogenèse des Acanthocéphales. 5. Flagellogenèse chez un eoacanthocephala: mise en place et désorganisation de l'axoneme spermatique. *Journal of Ultrastructure Research* **63**, 41–50.

Merritt, S. V. and Pratt, I. (1964). The life history of *Neoechinorhynchus rutili* and its development in the intermediate host (Acanthocephala: Neoechinorhynchidae). *Journal of Parasitology* 50, 394–400.

Meyer, A. (1928). Die Furchung nebst Eibildung, Reifung und Befruchtung des *Gigantorhynchus gigas*. Ein Beitrag zur Morphologie der Acanthocephalen. *Zoologische Jahrbucher. Anatomie und Ontogenie der Tiere* 50, 117–218.

Meyer, A. (1931). Urhautzelle, Hautbahn und plasmodiale Entwicklung der Larvae von *Neoechinorhynchus rutili* (Acanthocephala). *Zoologische Jahrbucher. Anatomie und Ontogenie der Tiere* 53, 103–126.

Meyer, A. (1933). *Acanthocephala. In* "Bronns Klassen und Ordnungen des Tierreichs", 4. Akademische Verlagsgesellschaft M. B. H., Leipzig.

Meyer, A. (1936). Die plasmodiale Entwicklung und Formbildung des Reisenkratzers (*Macracanthorhynchus hirudinaceus* (Pallas)). I. Teil. *Zoologische Jahrbucher. Anatomie und Ontogenie der Tiere* 62, 111–172.

Meyer, A. (1937). Die plasmodiale Entwicklung und Formbildung des Reisenkratzers (*Macracanthorhynchus hirudinaceus* (Pallas)). II. Teil. *Zoologische Jahrbucher. Anatomie und Ontogenie der Tiere* 63, 1–36.

Meyer, A. (1938a). Die plasmodiale Entwicklung und Formbildung des Reisenkratzers (*Macracanthorhynchus hirudinaceus* (Pallas)). III. Teil. *Zoologische Jahrbucher. Anatomie und Ontogenie der Tiere* 64, 131–197.

Meyer, A. (1938b). Die plasmodiale Entwicklung und Formbildung des Reisenkratzers (*Macracanthorhynchus hirudinaceus* (Pallas)). IV. Tiel. *Zoologische Jahrbucher. Anatomie und Ontogenie der Tiere* 64, 198–242.

Monné, L. and Hönig, G. (1954). On the embryonic envelopes of *Polymorphus botulus* and *P. minutus* (Acanthocephala). *Arkiv för Zoologi* Bd 7, nr 16, 257–260.

Moore, D. V. (1946). Studies on the life history and development of *Moniliformis dubius* Meyer, 1933. *Journal of Parasitology* 32, 257–276.

Moore, D. V. (1962). Morphology, life history, and development of the acanthocephalan *Mediorhynchus grandis* Van Cleave, 1916. *Journal of Parasitology* 48, 76–86.

Muzzall, P. M. and Bullock, W. L. (1978). Seasonal occurrence and host–parasite relationships of *Neoechinorhynchus saginatus* Van Cleave and Bangham 1949 in the fallfish, *Semotilus corporalis* (Mitchill). *Journal of Parasitology* 64, 860–865.

Nesheim, M. C., Crompton, D. W. T., Arnold, S. and Barnard, D. (1977). Dietary relations between *Moniliformis* (Acanthocephala) and laboratory rats. *Proceedings of the Royal Society of London Ser. B* 197, 363–383.

Nesheim, M. C., Crompton, D. W. T., Arnold, S. and Barnard, D. (1978). Host dietary starch and *Moniliformis* (Acanthocephala) in growing rats. *Proceedings of the Royal Society of London Ser. B* 202, 399–408.

Nicholas, W. L. (1967). The biology of the Acanthocephala. *Advances in Parasitology* 5, 205–246.

Nicholas, W. L. (1973). The biology of the Acanthocephala. *Advances in Parasitology* 11, 671–706.

Nicholas, W. L. and Grigg, H. (1965). The *in vitro* culture of *Moniliformis dubius* (Acanthocephala). *Experimental Parasitology* 16, 332–340.

Nicholas, W. L. and Hynes, H. B. N. (1958). Studies on *Polymorphus minutus* (Goeze, 1782) (Acanthocephala) as a parasite of the domestic duck. *Annals of Tropical Medicine and Parasitology* 52, 36–47.

Nicholas, W. L. and Hynes, H. B. N. (1963). The embryology of *Polymorphus minutus* (Acanthocephala). *Proceedings of the Zoological Society of London* 141, 791–801.

Nickol, B. B. (1969). Acanthocephala of Louisiana Picidae with description of a new species of *Mediorhynchus*. *Journal of Parasitology* **55**, 324–328.

Nickol, B. B. (1972). *Fessisentis*, a genus of acanthocephalans parasitic in North American poikilotherms. *Journal of Parasitology* **58**, 282–289.

Nickol, B. B. (1977). Life history and host specificity of *Mediorhynchus centrurorum* Nickol, 1969 (Acanthocephala: Gigantorhynchidae). *Journal of Parasitology* **63**, 104–111.

Nickol, B. B. and Heard, R. W. (1973). Host–parasite relationships of *Fessisentis necturorum* (Acanthocephala: Fessisentidae). *Proceedings of the Helminthological Society of Washington* **40**, 204–208.

Oetinger, D. F. and Nickol, B. B. (1974). A possible function of the fibrillar coat in *Acanthocephalus jacksoni* eggs. *Journal of Parasitology* **60**, 1055–1056.

Olson, R. E. and Pratt, I. (1971). The life cycle and larval development of *Echinorhynchus lageniformis* Ekbaum, 1938 (Acanthocephala: Echinorhynchidae). *Journal of Parasitology* **57**, 143–149.

Parenti, U., Antoniotti, M. L. and Beccio, C. (1965). Sex ratio and sex digamety in *Echinorhynchus truttae*. *Experientia* **21**, 657–658.

Parshad, V. R. and Guraya, S. S. (1977a). Correlative biochemical and histochemical observations on the lipids of *Centrorhynchus corvi* (Acanthocephala). *Annales de Biologie animale, Biochimie, Biophysique* **17**, 953–959.

Parshad, V. R. and Guraya, S. S. (1977b). Morphological and histochemical observations on the ovarian balls of *Centrorhynchus corvi* (Acanthocephala). *Parasitology* **74**, 243–253.

Parshad, V. R. and Guraya, S. S. (1978). Morphological and histochemical observations on oocyte atresia in *Centrorhynchus corvi* (Acanthocephala). *Parasitology* **77**, 133–138.

Parshad, V. R. and Guraya, S. S. (1979). Some observations on the testicular changes in an acanthocephalan, *Centrorhynchus corvi*, in natural infections of the crow, *Corvus splendens*. *International Journal of Invertebrate Reproduction* **1**, 262–266.

Parshad, V. R., Crompton, D. W. T. and Nesheim, M. C. (1980a). The growth of *Moniliformis* (Acanthocephala) in rats fed on diets containing various sugars. *Proceedings of the Royal Society of London Ser. B*, **209**, 299–315.

Parshad, V. R., Crompton, D. W. T. and Martin, J. (1980b). Observations on the surface morphology of the ovarian balls of *Moniliformis* (Acanthocephala). *Parasitology* **81**, 413–431

Petrochenko, V. I. (1956). "Acanthocephala of Domestic and Wild Animals, I." Akad. Nauk. SSSR, Moscow.

Petrochenko, V. I. (1958). "Acanthocephala of Domestic and Wild Animals, II." Akad, Nauk. SSSR, Moscow.

Podder, T. N. (1938). A new species of Acanthocephala, *Neoechinorhynchus topseyi* n.sp., from a Calcutta fish, *Polynemus heptadactylus* (Cuv. & Val.). *Parasitology* **30**, 171–175.

Rauther, M. (1930). Acanthocephala=Kratzwürmer. *In* "Handbuch der Zoologie", (W. Kukenthal, ed.), vol. II, pp. 3–5. Walter de Gruyter and Co., Berlin and Leipzig.

Robinson, E. S. (1964). Chromosome morphology and behaviour in *Macracanthorhynchus hirudinaceus*. *Journal of Parasitology* **50**, 694–697.

Robinson, E. S. (1965). The chromosomes of *Moniliformis dubius* (Acanthocephala). *Journal of Parasitology* **51**, 430–432.

Robinson, E. S. and Jones, A. W. (1971). *Moniliformis dubius*: X-irradiation and temperature effects on morphogenesis in *Periplaneta americana*. *Experimental Parasitology* **29**, 292–301.

Romanovski, A. B. (1964). The life-cycle of *Polymorphus minutus*. *Veterinariya* **42**, 40–41.

Rothman, A. H. and Fisher, F. M. (1964). Permeation of amino acids in *Moniliformis* and *Macracanthorhynchus* (Acanthocephala). *Journal of Parasitology* **50**, 410–414.

Schmidt, G. D. (1965a). *Polymorphus swartzi* sp.n., and other Acanthocephala of Alaskan ducks. *Journal of Parasitology* **51**, 809–813.

Schmidt, G. D. (1965b). *Corynosoma bipapillum* sp.n. from Bonaparte's gull *Larus philadelphia* in Alaska, with a note on *C. constrictum* Van Cleave, 1918. *Journal of Parasitology* **51**, 814–816.

Schmidt, G. D. (1969). Acanthocephala as agents of disease in wild mammals. *Wildlife Diseases* **53**, 10 pp. (microfiche).

Schmidt, G. D. (1973). Early embryology of the acanthocephalan *Mediorhynchus grandis* Van Cleave, 1916. *Transactions of the American Microscopical Society* **92**, 512–516.

Schmidt, G. D. and Huggins, E. J. (1973). Acanthocephala of South American fishes. Part 1, Eoacanthocephala. *Journal of Parasitology* **59**, 829–835.

Schmidt, G. D. and Kuntz, R. E. (1967). Revision of the Porrorchinae (Acanthocephala: Plagiorhynchidae) with descriptions of two new genera and three new species. *Journal of Parasitology* **53**, 130–141.

Schmidt, G. D. and Kuntz, R. E. (1969). *Centrorhynchus spilornae* sp.n. (Acanthocephala), and other Centrorhynchidae from the Far East. *Journal of Parasitology* **55**, 329–334.

Schmidt, G. D. and Neiland, K. A. (1966). Helminth fauna of Nicaragua. III. Some Acanthocephala of birds, including three new species of *Centrorhynchus*. *Journal of Parasitology* **52**, 739–745.

Schmidt, G. D. and Olsen, O. W. (1964). Life cycle and development of *Prosthorhynchus formosus* (Van Cleave, 1918) Travassos, 1926, an acanthocephalan parasite of birds. *Journal of Parasitology* **50**, 721–730.

Sinnott, E. W., Dunn, L. C. and Dobzhansky, T. (1958). "Principles of Genetics". McGraw-Hill, New York, Toronto and London.

Sita, E. (1949). The life-cycle of *Moniliformis moniliformis* (Bremser, 1811), Acanthocephala. *Current Science* **18**, 216–218.

Stranack, F. R. (1972). The fine structure of the acanthor shell of *Pomphorhynchus laevis* (Acanthocephala). *Parasitology* **64**, 187–190.

Uglem, G. L. (1972). The life cycle of *Neoechinorhynchus cristatus* Lynch, 1936 (Acanthocephala) with notes on the hatching of eggs. *Journal of Parasitology* **50**, 721–730.

Uglem, G. L. and Larson, O. R. (1969). The life history and larval development of *Neoechinorhynchus saginatus* Van Cleave and Bangham, 1949 (Acanthocephala: Neoechinorhynchidae). *Journal of Parasitology* **55**, 1212–1217.

Uznanski, R. L. and Nickol, B. B. (1976). Structure and function of the fibrillar coat of *Leptorhynchoides thecatus* eggs. *Journal of Parasitology* **62**, 569–573.

Van Cleave, H. J. (1914). Studies in cell constancy in the genus *Eorhynchus*. *Journal of Morphology* **25**, 253–298.

Van Cleave, H. J. (1920). Sexual dimorphism in the Acanthocephala. *Transactions of the Illinois Academy of Sciences* **13**, 280–292.

Van Cleave, H. J. (1940). The Acanthocephala collected by the Allan Hancock Pacific Expedition, 1934. *Allan Hancock Foundation: Publications*, Series 1, vol. 2, pp. 501–527.

Van Cleave, H. J. (1948). Expanding horizons in the recognition of a phylum. *Journal of Parasitology* **34**, 1–20.

Van Cleave, H. J. (1949a). Morphological and phylogenetic interpretations of the cement glands in the Acanthocephala. *Journal of Morphology* **84**, 427–457.

Van Cleave, H. J. (1949b). An instance of duplication of the cement glands in an acanthocephalan. *Proceedings of the Helminthological Society of Washington* **16**, 35–36.

Van Cleave, H. J. (1953). Acanthocephala of North American Mammals. *Illinois Biological Monographs* **23**, 1–179.

Verma, S. C. and Datta, M. N. (1929). Acanthocephala from Northern India. 1. A new genus *Acanthosentis* from a Calcutta fish. *Annals of Tropical Medicine and Parasitology* **23**, 483–500.

Walkey, M. (1967). The ecology of *Neoechinorhynchus rutili* (Miller). *Journal of Parasitology* **53**, 795–804.

Walton, A. C. (1959). Some parasites and their chromosomes. *Journal of Parasitology* **45**, 1–20.

Ward, H. L. and Nelson, D. R. (1967). Acanthocephala of the genus *Moniliformis* from rodents of Egypt with the description of a new species from Egyptian spiny mouse (*Acomys cahirinus*). *Journal of Parasitology* **53**, 150–156.

Warner, F. D. (1974). The fine structure of the ciliary and flagellar axoneme. *In* "Cilia and Flagella", (M. A. Sleigh, ed.), pp. 11–37. Academic Press, New York and London.

West, A. J. (1963). A preliminary investigation of the embryonic layers surrounding the acanthor of *Acanthocephalus jacksoni* Bullock, 1962 and *Echinorhynchus gadi* (Zoega) Muller, 1776. *Journal of Parasitology* **49**, (suppl.) 42–43.

West, A. J. (1964). The acanthor membranes of two species of Acanthocephala. *Journal of Parasitology* **50**, 731–734.

Whitfield, P. J. (1968). A histological description of the uterine bell of *Polymorphus minutus* (Acanthocephala). *Parasitology* **58**, 671–682.

Whitfield, P. J. (1969). "Studies on the Reproduction of Acanthocephala". PhD. Dissertation, University of Cambridge.

Whitfield, P. J. (1970). The egg sorting function of the uterine bell of *Polymorphus minutus* (Acanthocephala). *Parasitology* **61**, 111–126.

Whitfield, P. J. (1971a). Spermiogenesis and spermatozoan ultrastructure in *Polymorphus minutus* (Acanthocephala). *Parasitology* **62**, 415–430.

Whitfield, P. J. (1971b). Phylogenetic affinities of Acanthocephala: an assessment of ultrastructural evidence. *Parasitology* **63**, 49–58.

Whitfield, P. J. (1973). The egg envelopes of *Polymorphus minutus* (Acanthocephala). *Parasitology* **66**, 387–403.

Wright, R. D. (1971). The egg envelopes of *Moniliformis dubius*. *Journal of Parasitology* **57**, 122–131.

Yamaguti, S. (1939). Studies on the helminth fauna of Japan 29. Acanthocephala 2. *Japanese Journal of Zoology* **8**, 318–352.

Yamaguti, S. (1963). "Systema Helminthum. V. Acanthocephala". Interscience, John Wiley and Sons, New York and London.

Yamaguti, S. and Miyata, I. (1942). Uber die Entwicklungsgeschichte von *Moniliformis dubius* Meyer, 1933 (Acanthocephala), mit besonderer Berucksichtigung seiner Entwicklung im Zwischenwirt. Published by the authors, Kyoto, Japan.

Caryophyllidea (Cestoidea): Evolution and Classification

JOHN S. MACKIEWICZ

Department of Biological Sciences, State University of New York at Albany, Albany, 12222 New York, U.S.A.

I. INTRODUCTION

Since the discovery of *Caryophyllaeus* in 1781, there has been more speculation on the evolution of caryophyllid cestodes than on any other—and with good reason. With a monozoic body plan, a life cycle involving annelids, and boasting the only tapeworm (*Archigetes*) that can complete a life cycle in an invertebrate (tubificid annelid), caryophyllideans stand in sharp contrast to the far more numerous strobilate tapeworms that generally utilize arthropod intermediate hosts and lack examples with invertebrate definitive hosts. There is little speculation that cestodes arose from free-living flatworms, more on the identity of the specific ancestral flatworm and most on the phylogenetic relationships of caryophyllid cestodes to all others. Before these areas of speculation can be clarified, we must have answers to such questions as (a) what is the evolutionary significance of the monozoic body plan, (b) how do caryophyllids differ from strobilate cestodes, and (c) are caryophyllid cestodes ancestral to or secondarily evolved from strobilate cestodes? A consideration of these key questions forms the basis of this analytical review.

Much of the early literature on the biology, morphology and evolution of caryophyllid cestodes has been reviewed earlier (Mackiewicz, 1972). Since then there has been a number of significant papers dealing with cestode evolution, some of them breaking with traditional views. These recent papers (1971–1979) are reviewed after a consideration of the evolutionary significance of various aspects of caryophyllid biology and morphology. No new taxa nor nomenclatorial changes are proposed in this paper.

A. DEFINITIONS

Lack of uniform definitions can produce confusion in dealing with certain important concepts. Both Spengel (1905) and Rosen (1918) have used the term "monozootie" and "polyzootie" in describing whether a strobilate tapeworm is a single *individual* (monozootie) or a colony or collection of individuals (polyzootie). This same concept, with the terms "monozootic" and "polyzootic", has been used by Wardle and McLeod (1952) and Wardle *et al.* (1974) in separating the classes Cotyloda and Eucestoda. "Monozootie" or "monozootic" should not be confused with the very similar term "monozoic", which is used to refer to a tapeworm with a single set of reproductive organs. This meaning is unambiguous and, except for the above exceptions, has been used consistently and accurately throughout the literature on caryophyllid morphology. This is not true, however, for "progenesis" or "progenetic" and "neoteny" whose precise meaning is so crucial to an understanding of past and current interpretation of caryophyllid evolution. A summary of the problems associated with the definition of these terms was presented by Mackiewicz (1972).

For the purposes of this paper, I am utilizing the definitions from Gould (1977) who has not changed the original meaning of each word. *Progenesis* is (p. 485): "Paedomorphosis (retention of formerly juvenile characters by adult descendants) produced by precocious sexual maturation of an organism still in a morphologically juvenile stage". *Neoteny* is (p. 483): "Paedomorphosis (retention of formerly juvenile characters by adult descendants) produced by retardation of somatic development". The essential difference between the two is that there is an *acceleration* of maturation in progenesis; in neoteny, there is a *retardation* of somatic development. Both produce a mature individual with larval or juvenile features, but the mechanism for doing so is quite different. This differs from the definition of Smyth (1976) in which there is a continuum from progenesis, or the *advanced* development of genitalia in the larval or immature stage, to neoteny, the actual *sexual maturity* of a larval stage. Clearly it is difficult to distinguish between the two phenomena with this latter definition. In fact, one can argue that progenesis is incipient neoteny and the two are, or can be, the same phenomenon except for the time factor. The reader is here referred to Gould (1977) for an extensive analysis of the history and meaning of these often confused terms. I agree with Gould that it is the *process* rather than the *result* that is important in separating the two. In progenesis, it is the *immature* stage that is sexually mature; in neoteny, it is the *adult* stage that is sexually mature but juvenile characteristics have not kept the same pace and are retained in the true adult stage.

In my opinion, characters of parasites need not be purely morphological to qualify as "adult" or "juvenile", they can refer also to parts of the cycle. As Freeman (1973) points out, a characteristic of the adult stage of cestodes is their enteral habit, whereas the metacestode stage ("Juvenile stage") is parenteral. Hence, in *Archigetes* the coelom habitat as well as morphological features can be used to establish the developmental state of the individual.

B.   CESTODARIA

From time-to-time, caryophyllideans have been grouped with *Amphilina* and *Gyrocotyle* (Figs. 1a and 1b) in the subclass Cestodaria Monticelli, 1892. Variations on this theme continue to the present, despite the fact that the only basic similarity between caryophyllids and the two genera above is the monozoic body form. Features of life cycle, development and morphology of amphilinids and gyrocotylids are so different from caryophyllideans that to include all three groups together in one class, as recently proposed by Wardle *et al.* (1974), is to imply a close evolutionary relationship that has not been supported by any new evidence. On the contrary, workers who have recently studied evolutionary relationships of all monozoic cestodes (Freeman, 1973; Dubinina, 1974a, b; Malmberg, 1974; Stunkard, 1975) conclude that caryophyllideans are along different evolutionary lines from any of the other monozoic cestodes. Indeed, Dubinina (1974a) has erected the new class Amphilinoidea, considering it closer to the monogenetic trematodes than to the cestodes. All the data from earlier work and that since 1972 convince me that caryophyllids have evolved separately from amphilinids and gyrocotylids and therefore the latter two groups will be outside further discussions in this paper.

Also outside the coverage, are some other so-called monozoic forms. Because their morphology is quite unlike either *Amphilina* or *Gyrocotyle*, there seems no question whatsoever that the genus *Biporophyllaeus* Subramanian, 1939, is based on a detached proglottid. According to Joyeux and Baer (1961), it may be a Tetraphyllidean proglottid, whereas Yamaguti (1959) indicates that this genus, as well as *Anteropora* Subhapradha (1957), may be detached proglottids or hyperapolytic ones of Tetraphyllidean or Trypanorhynchidean cestodes. To these examples of isolated proglottids mistaken for cestodarians should be added the genus *Mastocembellophyllaeus* recently described by Shinde and Chincholikar (1977). Until the complete strobilae from which the detached proglottids came are described, it is pointless to make extensive systematic judgments, including synonomies, of any of these genera.

C. BIOLOGY OF CARYOPHYLLIDEANS

1. *General*

Caryophyllidea are monozic tapeworms parasitic in the intestine of freshwater fish, primarily of the orders Cypriniformes and Siluriformes. There are approximately 111 described species in 42 genera scattered in all zoogeographical regions with the largest number (46) in the nearctic and the smallest (1) in the neotropical region. There is a high degree of endemism with only two species, *Archigetes sieboldi* and *Glaridacris catostomi* being reported from more than one zoogeographical region. Fish become infected by eating tubificid worms (Oligochaeta) that harbor the metacestode stage. Aquatic oligocheates eat the operculate eggs that liberate a non-ciliated, six-hooked oncosphere which metamorphoses to the procercoid stage in the coelom or seminal vesicles of the tubificid worm. Except for some species of *Archigetes*, which may also mature in oligochaetes, the procercoid stage loses the cercomer on ingestion by the fish host and develops directly *in situ* into the plerocercoid-like adult stage. For more specific details of the life cycle, zoogeography and host–parasite relationships, see Mackiewicz (1972).

Caryophyllideans are not the only monozoic or unsegmented cestodes; nor are they the only ones that can mature in an invertebrate. In the small (less than six species) order Aporidea, species of the genera *Nematoparataenia* and *Apora* from duck and swans are monozoic, but share no other features with caryophyllids. *Gastrotaenia*, also of the order Aporidea, is not monozoic but lacks segments and like *Apora* is also found embedded in tissue (Willers and Olsen, 1969). According to Ginetsinskaya (1944), the lack of gonopores, characteristic of the order, and the subcutaneous habit can be considered aberrant characteristics indicative of neotenic forms.

The more common cestodes lacking external segmentation are the monotypic genera *Cyathocephalus* (Fig. 11b), *Spathebothrium* (Fig. 11c) and *Bothrimonus* (Fig. 11d; see Burt and Sandeman, 1969, for review of this genus) of the Spathebothridea, and the pseudophyllidean *Anantrum* Overstreet, 1968 (=*Acompsocephalum* Rees, 1969); *Cyathocephalus* is from salmonid and coregonid freshwater fishes, the others are from marine hosts. In *Anantrum*, the uterine pore and genital atrium are on opposite surfaces, and in *Bothrimonus* they are usually together and occasionally on alternate surfaces; in the other genera, the genital pores are also together but are normally on alternate surfaces. All have conspicuous internal proglottization and are thus unlike caryophyllids. The phylogenetic and systematic significance of the lack of segmentation and placement of genital pores has been discussed by Rees (1969).

Cyathocephalus, like *Archigetes*, is one of the few other cestodes that can mature in an invertebrate. On the basis of an extensive study of the development of *C. truncatus*, Wiśniewski (1932) concluded that this species was a neotenic plerocercoid. However, sexually mature forms were not found in the intermediate host, an amphipod crustacean *Gammarus*. More recently, Amin (1978) found 10 gravid procercoids of *C. truncatus* in the haemocoel of the large amphipod *Pontoporeia affinis*. Eggs were found in the haemocoel, but it was not clear whether they had come from a broken cuticular pouch (as in *Archigetes*) or from functional genital pores. Nevertheless, the fact that the procercoid stage was gravid is a clear indication of progenesis. In my opinion, the vertebrate phase of *Cyathocephalus* may also be a progenetic stage rather than an adult phase because it so closely resembles a plerocercoid with primary segmentation. Whether such progenetic procercoids can complete their cycle without a fish host is not known. In my opinion, it is quite possible that eggs liberated from a decaying *Pontoporeia* could be eaten by others to complete a one-host cycle; the success of such a cycle would depend on dispersion of eggs and the population density of the crustaceans. It appears to me, however, that the factors for success of a one-host cycle for *Cyathocephalus* are not as favourable as they have been for *Archigetes*.

2. Classification

The long and tortuous history of the classification of the monozoic tapeworms is one of the most complex of all the cestodes. As reviewed earlier (Mackiewicz, 1972), modern classification (since 1900) has regarded caryophyllids either as cestodarians, a family of the Pseudophyllidea, or as an independent order. Most helminthologists have abandoned the cestodarian status (but see Wardle et al., 1974) because of the absence of a 10-hooked, lycophora larva so characteristic of *Gyrocotyle* and *Amphilina* as well as morphological features such as the placement of the genital pores (Figs 1a and 1b). The Pseudophyllidean status, too, has lost considerable adherents for various reasons, among them the monozoic condition and realization that caryophyllids constitute a large group (over 100 species) that have oligochaetes as intermediate hosts. Since 1972 the separate ordinal status of the Caryophyllidea has been acknowledged by helminthologists who have had experience with the broad aspects of cestode classification (Freeman, 1973; Dubinina, 1974a; Wardle et al., 1974; Schmidt and Roberts, 1977) and is accepted here (see the discussion in Section VD).

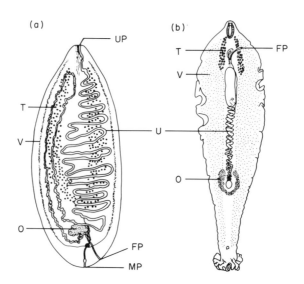

FIG. 1. (a) *Amphilina foliacea* (Rud.). (Adapted from Dubinina, 1974a.) Note position of uterine pore (UP), much removed from the female gonopore (FP) (compare with Fig. 11a). (b) *Gyrocotyle urna* (Grube and Wagener) Dorsal aspect. Male gonopore on venteral side (compare with Fig. 14). (Adapted from Lynch, 1945.)

3. *Intermediate hosts*

All evidence indicates that caryophyllids use only tubificid annelids as intermediate hosts (Table 1); in *Archigetes*, tubificids may also serve as definitive hosts. A list of the cestodes found in tubificid species may be found in Mackiewicz (1972) and Demshin (1975); additional records, primarily from the USSR, are in Grigelis (1972) and Dremkova (1974). With the studies on the life cycle of *Khawia sinensis* by Demshin (1977), *K. japonensis* by Demshin (1978), *Glaridacris vogei* by Williams (1978) and *Isoglaridacris wisconsinensis* by Williams (1980) the total number of caryophyllids experimentally infected in tubificids is now eleven. These data, in concert with the great number of records of naturally occurring infections, establish beyond doubt that caryophyllids do not need additional invertebrates, such as copepods or amphipods, to complete their development to the procercoid stage. Whether or not copepods or amphipods can be experimentally infected is not known, however.

All known life cycles of caryophyllideans involve aquatic oligochaetes of the families Naididae and Tubificidae. Among the 20 species recorded as intermediate hosts (Table 1) a number have a cosmopolitan distribution, and one, *L. hoffmeisteri*, is the most common and wide-spread tubificid known (Brinkhurst and Jamieson, 1971). An account of the biology of this species may be found in Kennedy (1966), and Brinkhurst and Jamieson (1971) have reviewed the aquatic oligochaetes of the world.

TABLE 1

Aquatic oligochaete hosts of caryophyllid cestodes

(Compiled from Demshin (1975), Dremkova (1974) and Grigelis (1972).)

Oligochaete	Number of cestode species
Class Oligochaeta	
Order Naidomorpha	
Family Naididae	
Stylaria lacustris (L.)	2
Dero digitata (Müller)	1
D. limosa Leidy	1
Unicinais unicnata (Oersted)	2
Ophidonais serpentina (Müller)	1
Family Tubificidae	
Euilyodrilus hammoniensis (Michaelsen)	3
Psamoryctes albicola (Michaelsen)	1
P. barbatus (Grube)	3
Peloscolex multisetosus (Smith)	2
Limnodrilus aurostriatus (Southern)	2
L. udekemianus Claparède	6
L. hoffmeisteri Claparède	14
L. claparedeanus Ratzel	5
L. cervix (Brinkhurst)	1
L. goti Hatai	1
L. willeyi Nomura	1
Tubifex tubifex (Müller)	8
T. templetoni Southern	4
T. barbatus Grube	1
T. hattai Nomura	1

In general, most species live in fresh water where they are a conspicuous part of the benthos, living in mud and sediment from which they extract organic matter (Stephenson, 1930). In addition to being an important but much underestimated source of food of fish (Kennedy, 1969), there are several important characteristics of tubificids that have a direct bearing on their role in the evolution of caryophyllids. Among these are (a) tolerance of low oxygen levels, (b) possessing a large coelom, and (c) having a relatively long life cycle.

The tolerance of tubificids to low oxygen tensions is well known. According to Palmer (1968), *T. tubifex* is a respiratory regulator down to a critical level of about 1·0 to 1·5% oxygen, below which its oxygen consumption drops off sharply. It is doubtful, however, that tubificids can respire anerobically for any length of time; it is possible, however, that they can be facultative anerobes (Brinkhurst and Jamieson, 1971). By being able to survive, in fact thrive, in an environment generally hostile to most other aquatic metazoans, tubificids remained as new, unexploited hosts (or intermediate hosts) for any newly evolving parasitic organism with stages that could also survive such environmental conditions.

The coelom is a large cavity that extends the length of the tubificid body, providing ample space for parasite (procercoid) development. For example, the length of the body for the three commonest intermediate hosts of caryophyllids is: *Limnodrilus hoffmeisteri*, 20–35 mm; *L. udekemianus*, 20–90 mm; and *Tubifex tubifex*, 20–200 mm. Of 83 species of caryophyllids, over 51% are less than 15 mm long as adults, with about 12 species less than 5 mm in length (Mackiewicz, 1972). Unlike the restricted space in the coelom of copepods or amphipods, that of tubificids can theoretically allow for growth beyond the normal metacestode stage, and in small species such as *Archigetes* this additional space, as well as a relatively long life span, may have been another of the factors leading to the evolution of progenesis in that genus. With the elimination of the definitive host and a shortening of the cycle, selection would favour such a progenetic form and eventually lead to the evolution of a progenetic species. Since tubificids had been established by the Permian period (Stephenson, 1930) more than 200 million years ago, there has been sufficient time for such associations to evolve.

According to Kennedy (1966), *L. hoffmeisteri* may take from one to two years to mature. Once infected with a caryophyllid, such a long life span greatly increases the period of infectivity thus compensating for a lowered egg output by the parasite (as compared with a polyzoic cestode). Any life cycle characteristic that increases the probability of parasite contact with the host would seem to have selective value.

By living in a habitat where oxygen often becomes a limiting factor, by feeding on mud and thereby allowing easy access within the oligochaete, by having a spacious coelom of the same shape and length as the worm itself and by living for more than a few months—these features collectively offered an extraordinary opportunity for exploitation by a newly evolving parasite. The wonder is that there have not been many other parasites as successful as the caryophyllids in evolving adaptations for cycles with the very common aquatic tubificid oligochaetes.

4. Definitive hosts

Vertebrate hosts of caryophyllids are exclusively freshwater fish although there are some scattered records, regarded here as accidental, from estuarine fishes such as *Pleuronectes* (Pleuronectidae), *Gobius* (Gobiidae) and *Zoarces* (Zoarcidae). Other rare and probably accidental hosts ingest infected tubificids while feeding on benthic organisms. Such host families as Clupeidae, Salmonidae and Percidae are of this type and serve to illustrate that a wide variety of hosts are exposed to and ingest infected tubificids, thus exerting selective pressure on cestode survival.

However, the dominant hosts are, by far, ostariophysan fishes (Table 2) with about 75% of the hosts from the cyprinoid families Cyprinidae (minnows) and Catostomidae (suckers). Next in importance are six siluriform families (catfish), which contain collectively about 17% of the hosts. These two orders contain over 90 % of caryophyllid hosts, a figure that strongly suggests a definite relationship (coevolution?) between host and parasite.

According to Greenwood *et al.* (1966), ostariophysan fish (a) consist of from 5000 to 6000 known species and thus constitute the major group of freshwater fish, (b) have some marine members (i.e. in the Plotosidae; see Table 2) and (c) are relatively primitive teleosts being placed near the base of a dendrogram showing evolutionary relationships of Division III (distinctively teleostean level ancestry) teleosts (Greenwood *et al.*, 1966, Fig. 1, p. 349). They are known from the Tertiary period. The Cyprinidae, the dominant hosts for caryophyllids, is the dominant family of freshwater fish in the world, with some 2000 known species and distributed on all continents except Australia and South America (Darlington, 1957). Catostomids, with less than 100 species, are found almost exclusively in North America with two species in Asia; the siluriform families are widely distributed in the Ethiopian and Oriental zoogeographical regions, with one (Plotosidae) from Australia.

TABLE 2

Zoogeographical distribution and principal families of freshwater fish hosts of caryophyllid cestodes

Hosts	Genera	Species	Zoogeographical region[1] (number of hosts)
Superorder Osteoglossomorpha			
Order Mormyriformes			
1. Mormyridae	1	1	E
Superorder Ostariophysi			
Order Cypriniformes			
Suborder Characoidei			
2. Characidae	1	1	E
Suborder Cyprinoidei			
3. Cyprinidae	40	51	P(37), N(10), O(3), E(1)
4. Catostomidae	9	25	N(25), P(1)
5. Cobitidae	3	4	P
Order Siluriformes			
6. Bagridae	3	3	E(2), O(1)
7. Clariidae	1	5	E(3), O(2)
8. Heteropneustidae	1	1	O
9. Mochokidae	1	4	E
10. Plotosidae	1	3	A
11. Schilbeidae	1	1	O
TOTALS	62	99	P(42), N(35), E(12), O(7), A(3)

[1] A, australian; E, ethiopian; N, nearctic; O, oriental; P, palearctic.

Although the hosts of the Caryophyllidea are a diverse group they have one common characteristic—similar feeding habits. For example, such species as *Mormyrus cashive* (Mormyridae), *Cyprinus carpio* (Cyprinidae), *Catostomus commersoni* (Catostomidae), *Clarias batrachus* (Clariidae), *Tandanus tandanus* (Plotosidae), *Synodontis schall* (Mochokidae) and *Heteropneustes fossilis* (Heteropneustidae) generally have benthic feeding habits. There can be little doubt that the feeding habits of the host have played a key role in the initial stages of the evolution of these cestodes by bringing tubificids and fish together, thus enabling a cycle to evolve. In my opinion, feeding habits are more important than the phylogenetic relationships of hosts *per se* when considering the evolutionary relationships of hosts to their parasites. *Esox* (pike, Esocidae), *Perca* (perch, Percidae) and *Micropterus* (bass, Centrarchidae) have not evolved as hosts of caryophyllids because they are primarily non-benthic feeders and not because they are phylogenetically more (or less) advanced than the cypriniformes, or even that they are non-ostariophysan. Whether caryophyllids are now physiologically incompatible with such hosts is problematical. However, judging from the extensive host list (Mackiewicz, 1972, Table V), which includes more than 104 species in six superorders, it is apparent that these tapeworms have a potentially wider host spectrum, limited to a great extent by the feeding habits of the host.

II. PERSPECTIVE

A. PROGENESIS

The essence of evolution involves the action of natural selection on the adaptations of organisms to their environment. These adaptations may involve morphology, physiology, development, behaviour, or the life cycle itself; in truth evolution is an interrelationship of all of these. But this fact does not mean that all adaptations have had an equal influence on the evolutionary direction of a group. So it is with the Caryophyllidea that progenetic development, much more widely spread in the Trematoda (Grabda-Kazubska, 1976), has been a key feature in their evolution.

Perhaps the best current treatment of progenesis and evolution is that of Gould (1977). The statement of Løvtrup (1978) notwithstanding, I believe the sections dealing with progenesis and life-history strategies provide a valid framework for viewing caryophyllid evolution. Those ideas or conclusions in Gould (1977) that have particular significance in understanding caryophyllid evolution are as follows.

1. *A primary variable in setting life-history strategy is the timing of maturation*

Progenesis is one important consequence of altering maturation time and thus itself becomes the object of natural selection. Rather than just a phenomenon that affects maturation only, it must be viewed in the context of the whole cycle in order to understand its role in the evolution of an organism. Features of a progenetic stage, such as morphology, site selection, or size, may in fact have no adaptive significance but be the normal consequence of progenetic development. Only by understanding the relationship between how progenesis affects survival and the developmental consequences of progenesis can one study the evolution of the Caryophyllidea. Without progenesis it is possible there would be no Caryophyllidea.

2. *Some adult features will accompany progenesis as a result of precocious maturation*

Maturation is an integration of complex physiological and developmental processes throughout a whole organism. One system does not complete its development without concommitant effects or changes in other systems. As a result of these interrelationships, a progenetic organism can be a mixture of "adult" and "juvenile" characters. Just as precocious maturation leads to the production of eggs, an adult developmental character, so too it may accelerate changes in other adult characters which may be more closely related to morphology. Viewed in this light, the presence of an adult morphological character (well-developed scolex, for example) becomes the *result* of progenesis and therefore is not evidence that the character is an adaptation *for* the progenetic stage.

3. *Progenesis plays a role in the rapid origin of higher taxa*

The conclusion is based on the answer to the following questions. What are the genetic consequences of progenesis? Or, what is the fate of the genes of the adult stage when that stage is not fully expressed in progenesis? Since evolution is basically a cytogenetic process (White, 1973) the fate of these "left over" genes may have a potentially profound effect on the evolution of an organism. By a process analogous to gene duplication that yields "extra" genetic material, Gould (1977) believes that the "unemployed" genes (he uses DeBeer's term) are transformable genes that are now available for experimental change. Progenesis is thus a mechanism that gives an organism unusual capacity to evolve rapidly in a new direction with very little genetic input. In theory, this mechanism allows for rapid evolution of new taxa because, though the gene transformations may be rare, few are needed to cause great changes. In speaking about origins of the Ctenophora, Gould (1977, p. 341) comments "Only one creative progenesis is required for the entire phylum". Surely the evolution of a monozoic tapeworm from a polyzoic one (or the reverse) is a relatively small and plausible step or series of steps when viewed against the enormous genetic potential and the millions of years of geological time available for experimentation and selection.

4. *Components of life-history strategies are adaptations selected by, and not merely consequences of, evolutionary process*

Timing of reproduction, fecundity, or other aspects of theoretical population ecology are integral aspects of the evolutionary process. Among other things, survival of a parasite depends ultimately on how well the whole cycle, *as a cycle*, has been selected for and adapted to the environment (or environments) in which it lives. Past studies on caryophyllid evolution have relied almost exclusively on classical evolutionary theory in which adaptations have been defined in terms of morphology or behaviour. Results, not process, have been elaborated. By not relating life cycle strategies to selective pressures, any scheme on the evolution of caryophyllids is lacking in justification and is little more than an exercise in speculation.

B. REPRODUCTIVE FITNESS

The monozoic body plan *per se* carries with it several important implications for the reproductive biology and evolution of caryophyllids. Consider for a moment that, by not having the capacity to strobilate, monozoic tapeworms are deprived of that feature which is characteristic of most helminth parasites—enhanced reproductive capacity. There can be little doubt that there is a net loss in reproductive capacity in monozoic forms compared with polyzoic cestodes actively producing proglottids. Furthermore, this net loss is not compensated for by asexually reproducing stages in an intermediate host (as in the Digenea, also monozoic) or by asexual reproduction in some metacestodes (i.e. *Echinococcus*). Thus, fitness in caryophyllids may be less dependent on the absolute reproductive capacity of the adult stages and more so on the interrelationships of their other aspects of biology; reproductive capacity, of course, remains an important aspect of that fitness. In my opinion, one of these aspects is the life cycle itself, especially the ecological relationships of the intermediate host, aquatic oligochaetes. Viewed in this perspective, one of the central themes directing the evolution of caryophyllids has been the natural selection of adaptations and loss of genetic capacity in the cyle that increases fitness in the absence of greatly enhanced reproductive potential.

In the final analysis, the factors that have contributed to the evolutionary success of caryophyllids should be related to reproductive fitness and to increasing the probability of completing a cycle. According to Fairbairn (1970), the major contributions to reproductive fitness are adaptations and loss of genetic capacity; adaptations are goal directed, the loss of genetic capacity is not. The difference between these two concepts is not always clear because, according to Fairbairn, the identification of an adaptation may be basically independent of the evolutionary history of a species, whereas the identification of the loss of genetic capacity always assumes that in the past such information was present. But to make this assumption, one must have a good understanding of phylogeny and evolution of the organism; unfortunately this is not true for caryophyllids. In the absence of a knowledge of the evolutionary history of caryophyllids and the functions or adaptive significance of characteristics, it is very difficult to assess accurately their relationship to reproductive fitness. Despite these severe constraints, the evolution of caryophyllids will not be fully elucidated unless each characteristic or life cycle feature is viewed in the total perspective of adaptations, loss of genetic capacity and reproductive fitness.

III. CHARACTERISTICS AND POSSIBLE EVOLUTIONARY SIGNIFICANCE

A. EXPERIMENTAL MONOZOIC FORMS

Caryophyllids are the only naturally occurring monozoic tapeworms with a hexacanth embryo (see Section IC1). Anomalies with a second set of reproductive organs on a lateral branch have been reported from *G. catostomi* by Mackiewicz (1978) and Williams (1979) and from *Penarchigetes* sp. (=*P. fessus*) by Mackiewicz (1978). As in *Cyathocephalus*, the dorsal and ventral surfaces of these individuals are not fixed; a reversal of the dorsoventral axis also appears to be the condition in *Taenia pisiformis* but not in *T. saginata*, *T. solium* or *D. latum* (Mueller, 1953). There is no evidence that any caryophyllid can produce multiple sets of reproductive organs other than by branching. There is some evidence that the reverse is true, namely that a monozoic condition can be derived from a polyzoic one.

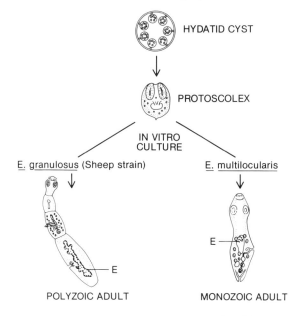

FIG. 2. Experimental production of monozoic adults of *Echinococcus multilocularis*. "Horse" and "cattle" strains of *E. granulosus* do not grow in culture. Monozoic adults appeared to have a full complement of reproductive organs but eggs (E) are unfertilized and do not embryonate. (Modified from Smyth and Davies (1975) and Smyth (1979).)

The experimental (*in vitro*) cultivation of sexually mature but sterile monozoic adults of *Echinococcus multilocularis*, less often *E. granulosus*, by Smyth and Davies (1975) and Smyth (1975) is the only direct experimental evidence that monozoic forms can be secondarily derived from a normally polyzoic species (Fig. 2). By using various media, these authors were able to produce easily and consistently up to 70 % monozoic forms (*E. multilocularis*) in culture, thus indicating that such production was not a rare, isolated event. Monozoic adults had a full complement of male and female reproductive organs but unlike polyzoic forms insemination had not taken place and eggs remained unfertilized and did not embryonate. Many questions regarding the process of strobilization, onset of sexuality in cestodes, and induction of monozoic differentiation in *E. multilocularis* are discussed by the authors. However, they state no conclusions regarding the evolutionary significance of this important discovery. It would appear that the production of monozoic forms would greatly strengthen the view that monozoic cestodes are secondarily derived from polyzoic ones. Until we are able to compare the cytodifferentiation in the neck of *Echinococcus* with that of any caryophyllid, such a conclusion is premature in my opinion.

A basic developmental difference between monozoic and polyzoic cestodes is that monozoic ones apparently lack the germinative region in the neck (Wiśniewski, 1930), so characteristic of polyzoic worms. It is proglottid production (or strobilization) and not segmentation that separates the two body forms. In *Cyathocephalus* (see Fig. 11b) and *Spathebothrium* (see Fig. 11c), for example, we have proglottidization but not segmentation. Because either polyzoic or monozoic forms can be produced, depending on the medium, it would appear that the germinal region in the neck of *Echinococcus* has been suppressed, rather than being absent. In my opinion, any cestode that is genetically polyzoic, regardless of the morphology, is basically a polyzoic tapeworm. The ultimate test that a cestode is basically monozoic or polyzoic must be decided at the genetic and cytodifferential level—not only from the morphology of the adult stage. It is quite clear from the outstanding work on *Echinococcus*, that through suitable manipulation in culture, a polyzoic worm may be genotypically polyzoic but phenotypically polyzoic or monozoic. True monozoic cestodes, as far as we know, are genotypically and phenotypically monozoic.

This is not to minimize the evolutionary importance of the experimentally produced forms. If segmentation can be suppressed *in vitro* surely it is possible that other basic developmental processes of polyzoic cestodes may have been altered through a long history of genetic experimentation and selection in the immensity of geological time. Whether similar alterations can be induced in metacestode or immature stages remains to be seen. It is important to bear in mind, however, that caryophyllids may not be *adult* polyzoic cestodes that have become monozoic through suppression of strobilization but that they are sexually precocious *immature* cestodes. Neither the procercoid nor plerocercoid stages are normally characterized by segmentation, hence their maturation by progenesis does not involve suppression but rather a truncation of ontogeny. Segmentation may occur in some plerocercoids in advanced stages of development, such as in the large plerocoids of *Ligula* or those of some *Diphyllobothrium* spp. whose plerocercoids form a primary strobila once ingested by a vertebrate (Freeman, 1973). Although a plerocercoid may not be segmented under the proper conditions, it may indeed form segments indicating that the capacity to segment is present. Progenesis may so truncate ontogeny that this capacity to segment is not expressed for lack of time rather than a true suppression of a capability that would ordinarily be expressed. Biologically and developmentally the monozoic *Echinococcus* can, therefore, be quite different from a caryophyllid though they both share the same morphological body plan.

Perhaps most important from an evolutionary point of view in the context of polyzoic *vs* monozoic, is that a cestode, as a biological entity or species, is composed of *all* life-cycle stages and not only the adult one. A change in only one stage cannot be interpreted as representing a basic change in the cestode as a life form. Without information on comparative cytodifferentiation between monozoic *Echinococcus* and caryophyllids, or whether or not the monozoic culture forms give rise to stages lacking asexual reproduction, a basic feature of the life cycle of caryophyllids, the monozoic *Echinococcus* should be regarded as aberrant or anomalous polyzoic tapeworms, comparable to polyradiate cestodes, and not monozoic cestodes at the same developmental or biological level as caryophyllids.

B. CALCAREOUS CORPUSCLE DISTRIBUTION

Calcareous corpuscles are a common characteristic of most cestodes (von Brand, 1973). They are a conspicuous element primarily of immature stages, being numerous and generally scattered randomly throughout the body. That there is indeed no discrete pattern to corpuscle distribution in the procercoid or plerocercoid stages can be verified by examining some representative illustrations of the following few species. Pseudophyllidea: *D. latum* (Rosen, 1918, Plate I, Fig. 2; Wardle and McLeod, 1952, Figs. 32F and 35), *D. norvegicum* (Vik, 1957, Plate I, Figs. 3 and 4), *Schistocephalus pungitti* (Dubinina, 1966, Figs. 82, 83), *Triaenophorus nodulosus* (Rosen, 1918, Plate II, Fig. 5), *Eubothrium salvelini* (Boyce, 1974, Figs. 10–12); Proteocephalidea: *Proteocephalus percae, P. macrocephalus* (Jarecka, 1960, Table I, Figs. 2 and 3; Table II, Fig. 5), *P. filicollis* (Freze, 1965, Figs. 39, 8–10); Cyclophyllidea: *Paruterina candelabraria* (Freeman, 1957, Plate I, Fig. 10); *Valipora campylancristrota, Neogryporhynchus cheilancristrotus, Paradilepis scolecina* (Kozicka, 1971, Figs. 1, 2 and 3) and *Ophiotaenia filaroides* (Mead and Olsen, 1971, Figs. 3 and 5).

Recently, Mackiewicz and Ehrenpris (1980) found that the distribution of calcareous corpuscles in some caryophyllids is unlike that in any other tapeworm. Briefly, they found that in *Glaridacris laruei* and *G. catostomi* the corpuscles are in discrete clusters that form a serially repeating pattern in two lateral dorsal and ventral rows (Figs. 3 and 4); there is little change in the number and distribution of clusters between small immature worms and much larger, mature ones; and clusters are lost in the posterior part of gravid worms but few in the organ-free neck region (Fig. 4d). Despite the basically different morphology of the two species, there was a mean of 22·4 cluster pairs (corresponding lateral clusters are considered a pair) with a range of 17–28 in 129 *G. laruei;* in 17 *G. catostomi* there was a mean of 23 with a range of 17–33 cluster pairs. These, as well as other data, indicated that morphology

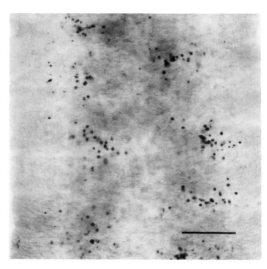

FIG. 3. Portion of *Glaridacris catostomi* illustrating the distribution of calcareous corpuscles that is strongly suggestive of cryptic segmentation. Approximately 33 lateral pairs of corpuscle clusters were present on the immature worm, 6 mm long. Note that each cluster does not always have a corresponding one on the opposite side. Corpuscles stained with silver nitrate; scale equals 0.1 mm. (From Mackiewicz and Ehrenpris, 1980; with permission of the editor of the *Proc. Helminthol. Soc. Wash.*)

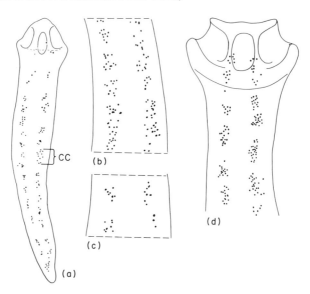

FIG. 4. Drawings of calcareous corpuscle distribution in *Glaridacris;* one side shown. (a) Immature *G. laruei*, corpuscle cluster (CC). (b) Immature *G. catostomi*. (c) Immature *G. laruei* showing four cluster pairs in neck region. (d) Scolex and neck of gravid *G. laruei*. (Adapted from Mackiewicz and Ehrenpris, 1980.)

per se is not the sole determinant of corpuscle distribution. They concluded that the loss of corpuscles in gravid worms was probably correlated with increased calcium utilization rather than an alteration of the basic pattern that appears to be a characteristic morphological feature of each species. No such pattern, however, was found in *Hunterella nodulosa* and *Monobothrium hunteri* where corpuscles occurred either in scattered clusters in the scolex or were randomly, but not homogeneously, distributed throughout the body of the worm.

Is there any evolutionary significance to the serial distribution of calcareous corpuscles in some caryophyllids? From a review of the diverse functions of corpuscles in invertebrates (Simkiss, 1976), trematodes (Erasmus and Davies, 1979) and cestodes (Chowdhury and DeRycke, 1977; Befus and Podesta, 1976) it does not seem unreasonable to assume that such ubiquitous structures have essentially similar functions in polyzoic and monozoic cestodes. If this assumption is correct, then it is probable that the unusual corpuscle *distribution* may be more related to some developmental or morphological feature *characteristic of the monozoic body plan*, than to a generalized physiological *function*.

After failing to correlate cluster distribution with any developmental or morphological feature, although correlation with the ganglia nodes on the lateral nerve cords of *G. laruei* was inconclusive, Mackiewicz and Ehrenpris (1980) concluded that the serially arranged clusters (1) "... are a form of cryptic segmentation that reflects differing physiological states in adjacent groups of cells", and (2) "... reflect a type of 'physiological segmentation' that preceded somatic segmentation and the formation of a strobila". In the first instance, there is no other corroborating evidence that supports this view; in the second, it assumes that caryophyllids had a strobila that was subsequently lost through the evolution of a progenetic cestode as Wiśniewski (1930) and Janicki (1930) proposed long ago. Yet one can also argue that such "physiological segmentation" had to precede somatic segmentation that has yet to evolve. As attractive as this alternative may be, it would suggest that caryophyllids were preadapted to be segmented. At present I see no genetic, physiological, or morphological basis for accepting this alternative view because there are no adaptive or selective pressures that would appear to have influenced the *expression* of this clustering characteristic *before* the appearance of the related strobilar morphology.

An alternative new interpretation related to strobilization is (3) that the corpuscle distribution is a vestige of the pattern in an ancestral *segmented* worm. In this case, we must assume that the genes for corpuscle distribution and function are linked. So vital are corpuscles for the life of a cestode that the genes for them (and their distribution) would be retained in the genome regardless of how many life cycle stages are dropped or added. Because the procercoid, plerocercoid and strobilate stages are in different hosts or different sites in the same host, thus being exposed to different physiological conditions, it is possible that there are separate genes regulating corpuscle distribution for each stage. The fact that there is no discernible pattern to corpuscle distribution in the immature stages of the cestodes mentioned above, would support this view because the genes regulating strobilization, and hence the segmental pattern of corpuscle distribution, have not been turned on. With progenesis, however, the genes for the sequential production of corpuscles (but not strobilization with which it was formerly related) in the *adult stage* are turned on and expressed, even though the strobilate stage no longer exists. Of possible significance in this regard is the fact that the segmental pattern is best developed in the neck region of *Glaridacris*; in polyzoic cestodes strobilization begins in the neck. If this analysis of events is accurate, then corpuscle distribution in some caryophyllids may be a true vestige of an *adult* character of polyzoic cestodes. The prospect that this may be the first visible indication of a masked gene related to segmentation in an ancestral polyzoic stage in the early evolution of caryophyllids is an exciting one.

C. INTRANUCLEAR GLYCOGEN VACUOLE IN VITELLINE CELLS

1. *General*

Electron microscope study of vitellogenesis of *Glaridacris catostomi* by Swiderski and Mackiewicz (1976a) has confirmed an earlier report (Mackiewicz, 1968) that the nuclei of mature vitelline cells have a single, large glycogen vacuole which serves as one of the food reserves in the egg. First appearing as beta glycogen particles, larger aggregates of alpha glycogen are soon formed, eventually fusing together to produce a very large, non-membrane-bound vacuole that displaces the nucleus to one side (Fig. 5a). Such a nuclear vacuole appears to be unique among cestodes, not being found in the vitelline cells of the proteocephalidean *Proteocephalus longicollis* by Swiderski et al. (1978), cyclophyllideans *Catenotaenia pusilla, Inermicapsifer madagascariensis* and *Hymenolepis diminuta* by Swiderski et al. (1970), tetraphyllid *Echeneibothrium beauchampi* by Mokhtar-Maamouri and Swiderski (1976a) and pseudophyllideans *Diphyllobothrium latum* by Schauinsland (1885, Plate 7, Fig. 2), *Cyathocephalus truncatus* by Wiśniewski (1932, Plate 13, Fig. 1), *Schistocephalus solidus* by Smyth (1956, Fig. 3), and *Bothriocephalus clavibothrium* by Swiderski and Mokhtar (1974). Vitelline cells of the proteocephalideans and tetraphyllids have glycogen and "yolk" or lipid as energy reserves whereas the caryophyllids and cyclophyllids have glycogen only (Mokhtar-Maamouri and Swiderski, 1976a). The presence of the nuclear vacuole is therefore not correlated with the absence of lipid in the vitelline cells; of greater importance is whether there are lipid reserves in the egg. Eggs of caryophyllideans and pseudophyllideans are quite similar to each other (Mackiewicz, 1968) each having an ovum surrounded by numerous vitelline cells and enclosed in a rigid capsule; however, caryophyllidean eggs have only glycogen as a reserve whereas those of the pseudophyllids have glycogen and lipid (Fig. 5b). The presence of both glycogen and fat in the eggs of pseudophyllideans is corroborated by the studies of Ginetsinskya et al. (1971) on *D. latum*. Her studies showed also that the eggs of the pseudophyllideans *Ligula columbi* and *Triaenophorus nodulosus* lacked glycogen but

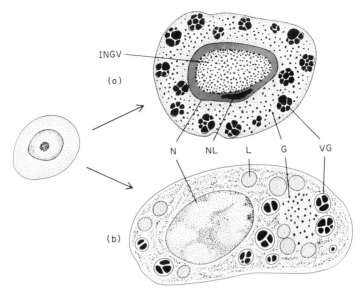

FIG. 5. Mature vitelline cells from electron microscope studies. (a) Caryophyllidea, *Glaridacris catostomi*; showing prominent intranuclear glycogen vacuole (INGV), glycogen in cytoplasm (G), and vitelline globules (VG). Note how the nucleolus (NL) is displaced to one side. (Adapted from Swiderski and Mackiewicz, 1976a.) (b) Pseudophyllidea, *Bothriocephalus clavibothrium*; Note absence of vacuole in nucleus (N) but presence of lipid (L) in cytoplasm. (Adapted from Swiderski and Mokhtar, 1974.) Both cells develop from a common type of gonial cell with a large nucleus and relatively little cytoplasm.

had lipid. In my opinion, this difference in energy reserve in the egg is correlated with the intranuclear glycogen vacuole, and may be of major consequence in the evolution of caryophyllid cestodes.

The presence of nuclei that synthesize and store glycogen as a *normal* cell function appears to be a unique phenomenon possibly confined to caryophyllideans. They have not been found in the vitelline cells of three polyopisthocotylean and one monopisthocotylean monogenea, nor in the digeneans *Fasciola hepatica* and *Schistosoma mansoni* (Halton *et al.*, 1974). No intranuclear vacuoles occur during vitellogenesis in the free-living triclad *Dugesia lugubris* according to Domenici and Gremigni (1974); nor are they known from any other turbellarian (J. B. Jennings, personal communication). I am unable to find any reference to such intranuclear glycogen vacuoles in any other invertebrate, although they have been described from vertebrate cells.

Among vertebrates, on the other hand, intranuclear glycogen has long been associated with pathological conditions particularly in the liver (Himes and Pollister, 1962). More recently, there has been considerable interest in studying the synthesis of intranuclear glycogen in tissue cultures of Ehrlich ascites cells (Zimmerman *et al.*, 1976). Tadpole liver cells appear to be one of the few places where intranuclear glycogen synthesis appears normally; however, in this case the occurrence is a sporadic phenomenon (usually less than 10% of the cells) and glycogen vacuoles disappear with metamorphosis of the tadpole (Himes and Pollister, 1962). Clearly, the formation of intranuclear glycogen vacuoles as a normal part of vitellogenesis in caryophyllid tapeworms is apparently a rare phenomenon in the animal kingdom.

2. Significance

What is the adaptive significance and possible selective value of the intranuclear glycogen vacuoles in the vitelline cells of these tapeworms? There are four attractive possibilities. The first is that, with an increase in glycogen, without a concommitant increase in the number of vitelline cells per egg or in egg size, the period of infectivity would be prolonged; such a characteristic would be selected for (Mackiewicz, 1968). An adaptation for prolonged periods of survival cannot be the entire picture because on an energy/weight ratio lipid would be a more favourable energy source than glycogen. If survival with respect to time were the only factor, then why would not lipid be a more favourable energy source than glycogen? The second is that, by having intranuclear *and* cytoplasmic glycogen, the energy reserves are partitioned and could be available at different times (Swiderski and Mackiewicz 1976a), much as in starving tadpoles, where the nuclear glycogen is utilized after cytoplasmic glycogen has been depleted (Himes and Pollister, 1962). Whether such partitioning takes place in normal embryogenesis or as a special adaptive feature to insure egg survival under adverse environmental conditions remains to be determined.

Regardless of the specific function, partitioning would allow for more efficient use of energy reserves, a characteristic that also would have selective value. The third possibility is that, since vitelline cells, and hence eggs, lack "yolk" or lipid energy reserves, the increase in glycogen per cell may help to compensate (balance?) this energy loss without increasing the number of cells per egg. Unfortunately, there are no comparative data on the calorific values of caryophyllid and other comparable eggs, but with lipid. If we assume that the calorific values were somewhat equal then the increased glycogen would enable the species to survive in the absence of the more common lipid–glycogen reserve.

Before going further, it is important to examine some of the glycogen–lipid energy relationships in cestodes. As Calow and Jennings (1974) have found, free-living platyhelminths generally have an energy source rich in lipids whereas entosymbionts generally are rich in glycogen. Reasons for this difference are related to the stable food source in the gut, which removes the need for long-term storage, and the energetics of high fecundity. More recently, Jennings and Calow (1975) elaborated on their hypothesis stating that the large quantities of glycogen are an adaptation for the high fecundity which is an "automatic consequence" of the nutrient-rich gut. Formerly, it was believed, however, that the high glycogen content of parasitic worms was primarily an adaptation, or pre-adaptation (Jennings, 1973), to the low or variable oxygen tensions of the gut. Regardless of whether or not the high glycogen is an adaptation for, or a consequence of, the nutrient-rich gut, the low oxygen tensions necessitate that glycogen and not lipid be used as an energy source. In caryophyllid biology, I believe that this same glycogen–low oxygen relationship may not apply directly to the fecundity question (because no proglottids are formed) but to the ecology of the eggs. Considering that glycogen is heavier than lipid and that unlike lipid it can be utilized under anerobic conditions, these two characteristics are consistent with a life cycle that has a benthic intermediate host (see Section IC3).

As a result of these characteristics, there is a fourth possibility, the most important one for the adaptive significance of intranuclear glycogen. As a consequence of having additional glycogen in the egg to compensate (balance?) for the lack of lipid, the caryophyllid egg is better adapted *to sink* and *to survive* in a habitat (mud) with little or variable oxygen tension and as a esult the organism is able to make an evolutionary "breakthrough"—*that of enabling viable eggs to be exposed to, and exploited by, a new potential intermediate host, tubificid annelids.* As so elegantly demonstrated by Jarecka (1961), the eggs of tapeworms with aquatic cycles are adapted in various ways that greatly enhance the probability of their being eaten by intermediate hosts such as Copepoda, Ostracoda, Cladocera, Amphipoda and Oligochaeta. Without doubt, the caryophyllids have been the most successful helminths in exploiting these annelid hosts, who live in a physiologically hostile environment. According to Demshin (1975, pp. 158–161) the only other cestodes that have oligochaetes as intermediate hosts are *Hymenolepis moghensis, Paricterotaenia* (=*Sacciuterina*) *stellifera, Haploparaxis* (=*Monocholepis*) *dujardini, Aploparakis filum* and *A. furcigera*, all cyclophyllideans of birds. The eggs of *A. furcigera* also sink, being large and in small, relatively heavy packets (Jarecka, 1961). Nematodes have been less successful, with only two species recorded (Demshin, 1975). Even the marine polychaeta, other benthic annelids, rarely serve as intermediate hosts of cestodes (Margolis, 1971). From the very beginning, the absence of much competition for tubificid intermediate hosts allowed caryophyllids successfully to incorporate them into a cycle that also involved benthic feeding fish, thus establishing the basic caryophyllid cycle.

If the preceding analysis of the adaptive significance of intranuclear glycogen vacuoles is correct, it is not surprising that other cestodes have not been able to exploit tubificids because of the limited occurrence of such vacuoles. Furthermore, if the nuclear vacuoles are primarily an adaptation for egg survival and development, and not for some physiological functions of the mature worm, there is little reason to search for them in free-living or symbiotic turbellarians as possible clues to ancestral stock. The apparent absence of intranuclear vacuoles in the turbellaria is, therefore, of little evolutionary significance.

D.   CYTOLOGY

Since 1972, our knowledge of the chromosome numbers of the Caryophyllidea has increased considerably (Table 3), largely because the chromosomes are common in testes squashes and they can be readily stained with leucobasic

160   JOHN S. MACKIEWICZ

TABLE 3

Chromosome numbers of caryophyllid cestodes

Species	2n	References
Caryophyllaeidae		
Archigetes sp.	18	Motomura (1929)
Hunt ralla nodulosa Mackiewicz	14	Mackiewicz and Jones,
and McCrae		(1969), Grey (1979)
Glaridacris laruei (Lamont)	16	Grey and Mackiewicz (1974)
	16, 18; 24–27 (3n)	Grey (1979)
G. confusus Hunter	16	Grey (1979)
G. catostomi Cooper	20; 30 (3n)	Grey and Mackiewicz (1980)
G. vogei Mackiewicz	20	Grey (1979)
Monobothrium hunteri Mackiewicz	20	Grey (1979)
Biacetabulum biloculoides		
Mackiewicz and McCrae	20	Grey (1979)
Isoglaridacris folius Fredrickson		
and Ulmer	18	Grey (1979)
I. jonesi Mackiewicz	18	Grey (1979)
I. bulbocirrus Mackiewicz	18; 27 (3n)	Grey (1979)
Caryophyllaeus laticeps (Pallas)	30 (3n)?	Grey (1979)
Lytocestidae		
Atractolytocestus huronensis	24 (3n)	Jones and Mackiewicz (1969),
Anthony		Grey (1979)
Lytocestus indicus (Moghe)	16	Vijayaraghavan and
		Subramanyam (1977)
Khawia iowensis Calentine and		
Ulmer	16	Grey (1979)
K. rossittensis Szidat	16	Grey (1979)
Notolytocestus minor Johnson		
and Muirhead	12	Grey (1979)
Caryoaustralus sprenti Mackiewicz		
and Blair[1]	6	Grey (1979)
Capingentidae		
Capingens singularis Hunter	14	Grey (1979)
Balanotaeniidae		
Balanotaenia bancrofti Johnson	14	Grey (1979)

[1] Listed as "gen. et sp.n. from Australia", recently described by Mackiewicz and Blair (1980).

fuchsin. Unlike chromosomes from other cestodes those of caryophyllids are quite large; for example, the largest (in μm) for several species are: *Hunterella nodulosa*, 6–8; *Glaridacris laruei* (diploid), 7–12; and *G. catostomi* (triploid) 6–8. For the first time it is now possible to make accurate idiograms and comparative studies of cestode karyotypes, techniques commonly used in contemporary systematic studies of other organisms. With refined techniques such as banding, it may be possible also to get precise information on translocations, inversions, deletions and duplication on chromosomes and thus learn something of the cytogenetics involved in cestode speciation. With this combination of desirable cytological characteristics, it is little wonder that caryophyllids have been designated the *Drosophila* of tapeworms (Mackiewicz, 1976).

Although only 18 species have been studied cytologically (Table 3) there are some patterns emerging that may aid in understanding the evolution and systematics of this group of cestodes. For example, there is a greater range in chromosome numbers (6 to 20) than has been found in any other order. The Lytocestidae have the lowest number; the genus *Glaridacris* has more than one number but *Isoglaridacris* has only one; and some species, such as *Glaridacris laruei*, have at least two cytologically distinct diploid populations. Unlike the Pseudophyllidea where $2n=18$ for four species of *Diphyllobothrium*, the wide range in chromosome numbers for caryophyllids suggests that the evolution of the Caryophyllidea was accompanied by various changes in chromosome number and other structural rearrangements of the karyotype. Clearly, the Caryophyllidea provide a rare opportunity to study the cytotaxonomical relationship within a cestode group. Cytological studies are needed to better understand the systematic relationships of genera to each other and to intepret the difficult questions dealing with intra- and interspecific variation.

However, the greatest discovery by far is that of polyploidy (triploidy), to date not described from any other cestode. Chromosome numbers are known for approximately 53 species of strobilate tapeworms. Polyploidy in *Atractolytocestus huronensis* and *Glaridacris catostomi* has been described in detail by Jones and Mackiewicz (1969) and Grey and Mackiewicz (1980); further examples (Grey, 1979) will be described in subsequent publications. In both cases of triploidy, spermatogenesis was abnormal to the point of complete failure and parthenogenesis was presumed to occur because eggs remained unfertilized. This cytological evidence, as well as the lack of any for hybridization, indicates that these triploids are autoploids. Except for *A. huronensis*, which may have *Markevischia sagittata* (chromosome number unknown) from the Amur carp in North Asia as the diploid parent or biotype (Jones and Mackiewicz, 1969), diploid and triploid forms have been demonstrated in the other three species, a fact that has an important bearing on the evolution and speciation of caryophyllids.

The role of polyploidy in evolution and speciation in animals has been studied much less than in plants; for reviews see Mayr (1963), White (1973, 1978) and Jackson (1976). As in caryophyllids, most cases of polyploidy in animals are associated with parthenogenesis because of the adverse effects of ploidy on meiosis. It appears, however, that polyploidy, despite its incompatibility with sexual reproduction, has selective advantages. According to Jackson (1976), polyploids have a greater ecological amplitude that allows them to better exploit new environments, even though they have the identical genes of a diploid. In many organisms, where survival of widely dispersed progeny is a key factor in species survival, the ability to survive in new environments would clearly have adaptive value. However, in obligate parasites

with complex life cycles involving more than one host, increased (physiological) amplitude must involve all stages in the cycle in order to maintain
the complex host–parasite interrelationships in the cycle. Unless this is done,
the cycle would fail and, with it, the species. In such highly specialized parasites as cestodes, survival depends not so much on being able to radiate into
new hosts but in evolving ways of consistently reaching the same or narrow
group of hosts. The hallmark of cestodes is specificity, not adaptability. An
increase in the number of eggs or reproductive potential has a much greater
adaptive and selective value than has the increased ecological amplitude of
adults. Indeed, one of the chief diagnostic features of cestodes is the chain
of proglottids which provides such large numbers of eggs. Perhaps polyploidy
is so rare (absent?) in strobilate tapeworms because an increase in the ecological amplitude of any stage would so disrupt the cycle that such stages would
be selected against. On the other hand, the ability to survive in diverse
ecological situations may have played an important role in the early evolution
of parasite species. How such a possible increase in ecological amplitude
resulting from polyploidy might be applied to *A. huronensis* in the carp,
Cyprinus carpio, was discussed at some length by Jones and Mackiewicz
(1969).

Another closely related "benefit" of polyploidy (also based on data from
free-living organisms) is the ability of polyploids to regulate some sets of
genes whereas others are expressed totally. The net effect is to give certain
types of polyploids, "exceptional genetically based amplitude with which
they should be able to move into new habitats with requirements beyond
their progenitors capabilities" (Jackson, 1976, p. 223). Since the functional
aspects of this "benefit" would serve also to increase the ecological amplitude
of the species, it would probably not be adaptive for cestodes for the reasons
discussed above. Furthermore, the triploid caryophyllids are not found in a
wider variety of hosts, a fact that tends to support the view that polyploidy
has not led to a wider host spectrum. Regrettably, there are no data on the
comparative distribution of polyploids and triploids in the intestine. Nor do
we know if polyploids in single-species infections have larger populations
because of an increased capability to extend beyond their normal sites in
the intestine; or if polyploids have a competitive advantage for available sites
in multiple infections (two to four species), which may be as high as 29 % in
some regions (Mackiewicz *et al.*, 1972).

A final effect of polyploidy is to increase the size of the organism. In the
only species studied in detail, diploid and triploid individuals of *Glaridacris
catostomi* were very similar in size (Grey and Mackiewicz, 1980). However,
though not significantly different in size, ten mature triploids had a mean
size of 23·2 mm (range, 10–47 mm) whereas ten diploids had a mean size
of 17·5 mm (range, 11·5–24 mm) (A. J. Grey, personal communication). It is
not unreasonable to assume that a larger parasite would have a greater
capacity to store eggs but whether there would be an absolute increase in
egg production is not known. If there is an increase in egg production (see
Section IIIE) correlated with increased worm size, then even though the

increase in body size between diploid and triploids is not statistically significant, at the population level this difference might result in enough eggs
being liberated by the polyploids to have them selected for over the diploid
ones. In the Bozenkill river, only triploid *G. catostomi* were found, indicating
that the polyploids had indeed been selected for but the exact reason for this
dominance has yet to be explained (Grey and Mackiewicz, 1980).

Without more information on reproductive potential, site preference and
population biology of diploids and polyploids in the same species, it is difficult
to assess the evolutionary significance of polyploidy in the caryophyllids. Of
course the most immediate effect of triploidy has been to alter the reproductive biology of the individual from sexual to asexual reproduction
(parthenogenesis). The net effect of this change is to have the potential for
establishing populations, as in the Bozenkill river, with virtually no gene
flow and essentially little further evolution. Combined with the consequences
of parthenogenesis, discussed in the next section, possible increased egg
production (correlated with increased body size) may serve to give polyploids
selective advantage over their diploid parent species. If this assumption is
true, then it is not unreasonable to expect polyploidy to occur with increasing
frequency in caryophyllideans.

E. PARTHENOGENESIS

Parthenogenetic reproduction among tapeworms is rare (Cable, 1971).
Ilisha parthenogenetica (of unknown systematic status), known only from
plerocercoids in the mesentery and liver of the Indian shad *Hilsa ilisha*, was
originally thought to be a parthenogenetic adult (Southwell and Prashad,
1918) but later Southwell and Prashad (1923) thought it was reproduction by
a plerocercoid in a manner similar to the production of germ balls by trematodes. According to Wardle and McLeod (1952), however, *Ilisha* reproduced by endogenous budding. A report (Coil, 1970) that the dioecious
cyclophyllidean *Gyrocoelia* is parthenogenetic was subsequently corrected
(Coil, 1972).

Parthenogenesis in the Cestoidea would appear then to be confined to the
Caryophyllidea. First described from the triploid *Atractolytocestus huronensis*
by Jones and Mackiewicz (1969) it has since been found in three other species.
Study of spermatogenesis of the triploid "race" of *Glaridacris catostomi*
reveals that functional sperm are not produced and parthenogenesis is presumed to occur (Grey and Mackiewicz, 1980). Similar conditions appear to
be true (Grey, 1979) in two other cases of polyploidy (Table 3), *G. laruei* and
Isoglaridacris bulbocirrus. Since the effect of parthenogenesis is to produce
cestodes that are essentially functional females one could use the term
"thelytoky" rather than parthenogenesis. It is clear that without parthenogenesis none of the sterile polyploids would survive. Except for *A. huronensis*,
which apparently has no diploid forms in North America, the other three
species have such diploid forms, a condition that raises important problems
regarding the systematic status of diploids and triploids of the same species.
Some of these problems have been addressed by Enghoff (1976) and White
(1978).

No less important are the evolutionary implications with populations of parthenogenetically produced clones. The evolutionary significance of parthenogenesis has been discussed at length by many workers, among them Suomalainen (1962), Suomalainen et' al. (1976), Mayr (1963), Tomlinson (1966), White (1973, 1978) and Maynard Smith (1978). Except for Price (1977), who has an illuminating discussion of parthenogenesis and parasite evolution, the other treatments generally deal with biparental, non-parasitic species. The type of parthenogenesis may determine its evolutionary significance. Oogenesis has not been studied in any of the triploid caryophyllids hence it is not known whether the parthenogenesis is of the automictic or apomictic type, the latter being most common (Suomalainen et al., 1976). If apomictic, there are no new gene combinations, except by mutation, and there is a greater chance of giving rise to a stable parthenogenetic form; the offspring retain the genetic make up of the parent worm. In ecologically diverse environments, the lack of genetic diversity resulting from this genetic stability would have an adverse effect on the evolution of a species. However, the disadvantage of having genetically uniform progeny is lowered for an intestinal parasite because of a highly stable and predictable environment provided by the homoeostasis of the host (Price, 1977). Such genetic stability (resulting from parthenogenesis) may, in fact, be selected for if particularly adaptive gene combinations are fixed in homozygous individuals, especially, according to Price (1977), those gene combinations that may be essential in the coevolutionary tracking of the host system. Because of their hermaphrodism, strobilate structure, isolation in hosts and ease of self-fertilization, tapeworms are considered largely homozygous (Jones, 1967; Logachev, 1970). If the maintenance of genetic stability is an adaptive feature for some intestinal parasites such as cestodes, why then should parthenogenesis, one way of preserving genetic stability (another would be selfing, undoubtedly a common phenomenon in strobilate tapeworms), be much more common in (restricted to?) the Caryophyllidea than in other cestodes?

A possible answer may be related to the monozoic body plan. It is obvious that strobilization greatly increases the reproductive potential of polyzoic cestodes; without such an increase it is doubtful that polyzoic cestodes would have been successful parasites. Caryophyllids, however, lack the reproductive advantage associated with strobilization yet are also successful parasites. How does one resolve this apparent contradiction? Unlike trematodes, also monozoic, and small tapeworms with few proglottids (e.g., Echinococcus) where the limited reproductive capacities in their adult stages have been greatly supplemented by a much increased asexual reproductive capacity in other stages, caryophyllids lack all traces of asexual reproduction in any part of their cycle and have the basic reproductive formula: 1 egg=1 adult. Clearly, any adaptation or change in the reproductive biology that would increase egg production or egg survival would have obvious selective value. Unfortunately, we do not have any data on either of these two important aspects of caryophyllid biology, but on theoretical grounds it is possible that through parthenogenesis, the energy normally devoted to sperm and egg production could all be used for egg production, thus effectively increasing the reproductive potential. Through parthenogenesis, such individuals would have greater fitness under non-stress conditions and also when the density of tubificids or fish is low. By having sexually and parthenogenetically reproducing individuals in the same species, and thus having the benefits of both modes of reproduction, the species would have a greater fitness than could be achieved with only one form of reproduction. Whether this reproductive polymorphism is wide-spread (we know of it in three species so far) remains to be seen. In my opinion, however, parthenogenesis, so wide-spread in parasitic and non-parasitic arthropods (Price, 1977) and found in numerous other animals including vertebrates, is an important, and perhaps even common, adaptation preventing sterility (resulting from polyploidy) and at the same time may increase the reproductive potential of caryophyllids.

F. SPERM MORPHOLOGY

Fine structure of sperm is known for only one species—*Glaridacris catostomi*. The spermatozoon is a very elongate, filiform structures about 260 μm long consisting of a body portion and a single axoneme (flagellum) with a 9+1 structure (Swiderski and Mackiewicz, 1976b). As in other cestode sperm, an acrosome and mitochondria are absent. Paradoxically, the presence of a single axoneme places the Caryophyllidea together with the Cyclophyllidea, anoplocephalids and some tetraphyllids (Table 4; Figs 6a–g) with which they share few other characteristics, and not the Pseudophyllidea, that have two axonemes, with which they are most often associated. Although sperm morphology has been used to assess phylogenetic and taxonomical

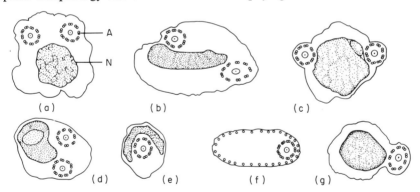

FIG. 6. Diagrammatic drawings from electron micrographs of cestode spermatozoa. (a) Trypanorhyncha: *Lacistorhynchus tenuis* (from Swiderski, 1976), axoneme (A), nucleus (N). (b) Proteocephalidea: *Proteocephalus longiocollis* (from Swiderski and Eklu-Natey, 1978). (c) Pseudophyllidea: *Diphyllobothrium latum* (from von Bonsdorff and Telkka, 1965). (d) Tetraphyllidea: *Acanthobothrium filicolle* (from Mokhtar-Maamouri and Swiderski, 1975). (e) *Echeneibothrium beauchampi* (from Mokhtar-Maamouri and Swiderski, 1976b). (f) Cyclophyllidea: *Hymenolepis diminuta* (from Sun, 1972). (g) Caryophyllidea: *Glaridacris catostomi* (from Swiderski and Mackiewicz, 1976b).

TABLE 4
Axoneme number in cestode and some turbellarian spermatozoa

Organism	Reference
ONE AXONEME (9+1)	
CESTOIDEA	
Caryophillidea	
Glaridacris catostomi	Swiderski and Mackiewicz (1976b)
Cyclophyllidea	
Hymenolepis nana	Rosario (1964)
Echinococcus granulosus	Morseth (1969)
H. diminuta	Rosario (1964), Lumsden (1965), Sun (1972)
H. microstoma	Swiderski (1970)
Catenataenia pusilla	Swiderski (1970)
Inermicapsifer madagascariensis	Swiderski (1970)
Taenia hydatigena	Featherston (1971)
Anoplocephalidea	
Monieza expansa	Swiderski (1968)
Tetraphyllidea	
Echeneibothrium beauchampi	Mokhtar-Maamouri and Swiderski (1976b)
TWO AXONEMES (9+1), SINGLE AXIAL FILAMENT	
CESTOIDEA	
Pseudophyllidea	
Diphyllobothrium latum	von Bonsdorff and Telkka (1965)
Tetraphyllidea	
Acanthobothrium filicolle	Mokhtar-Maamouri and Swiderski (1975)
Onchobothrium uncinatum	Mokhtar-Maamouri and Swiderski (1975)
Trypanorhyncha	
Lacistorhynchus tenuis	Swiderski (1976)
TURBELLARIA	
Rhabdocoela	
Kalyptorhynchia	Hendelberg (1969, 1974, 1977)
Acoela	Hendelberg (1969, 1974, 1977)
TWO AXONEMES (9+1), TWO AXIAL FILAMENTS	
CESTOIDEA	
Proteocephalidea	
Proteocephalus longicollis	Swiderski and Eklu-Natey (1978)
TURBELLARIA	
Rhabdocoela	
Dalyellioida	Hendelberg (1969, 1977)
Typhloplanoida	Hendelberg (1969, 1977)
ONE AXONEME (9+2)	
TURBELLARIA	
Nemertodermatida	Tyler and Rieger (1975), Hendelberg (1977)
Rhabdocoela	
Kalyptorhynchia (?)	Hendelberg (1977)

relationships in the turbellaria by Hendelberg (1974, 1977), it is clear that great caution should be exercised when using sperm morphology as an indicator of phylogenetic relationships within the cestoidea. Yet when one considers that the primitive metazoan sperm has a single axoneme (Franzen, 1956) like that of the Nemertodermatida, which have been separated from the Acoela and are placed near the base of the phylogenetic tree of platyhelminth relationships by Hendelberg (1977), then the occurrence of a single axoneme in some cestodes (i.e., caryophyllids) may not be due to reduction, as suggested by Hendelberg (1970), but rather evidence of a close relationship to primitive turbellarians. If this is the case, how does one reconcile the presence of one axoneme (Fig. 6g) in the sperm of acknowledged primitive cestodes (Caryophyllidea) as well (Fig. 6f) as highly advanced ones (Cyclophyllidea)? Without some knowledge of the fertilization biology of concerned groups and the factors that influence sperm motility it will be difficult to answer this question.

G.   VITAMIN B_{12}

The presence of high concentrations of vitamin B_{12} (cyanocobalamin) in *D. latum* and its relationship to pernicious anaemia in humans is well known (Tötterman, 1976). High concentrations have also been recorded from other pseudophyllidean cestodes such as the plerocercoids of *Ligula* sp. and *Spirometra mansonoides*, where the values have been, respectively, between 200–600 and 100–600 µg B_{12} per 100 g dry weight (Weinstein and Mueller, 1970). Rausch *et al.* (1967) reported a value of 2·6 µg B_{12} per g dry weight for adults of *Schistocephalus solidus*. In contrast to the high concentrations in pseudophyllidean cestodes, no detectable amounts of vitamin B_{12} have been found in cyclophyllidean cestodes, namely *H. diminuta* and *Taenia taeniaeformis*, using microbiological assays (Weinstein and Mueller, 1970; Tkachuck *et al.*, 1976).

Since first reporting a pink colour for some specimens of *Biacetabulum infrequens* and *Glaridacris laruei* (Mackiewicz, 1972), assays for vitamin B_{12} have been done on two species. Radioisotope assay of adult *Hunterella nodulosa* and *Glaridacris laruei* from *Catostomus commersoni* yielded values of vitamin B_{12} per 100 g dry weight of 248·9 and 1228·8 µg respectively. Although these results are preliminary, they establish that caryophyllidean cestodes, like pseudophyllideans studied thus far, have high concentrations of vitamin B_{12}. Apparently cestodes with vitamin B_{12} are able to form propionate from succinate, whereas those without cyanocobalamin cannot. The implications for energy metabolism for these pathways in cestodes have been discussed by Tkachuck *et al.* (1977).

Can one attribute any selective advantage to cestodes with the capacity to accumulate large amounts of vitamin B_{12}? It is already known that cestodes that accumulate vitamin B_{12} also have propionate as a product of anerobic energy metabolism (Tkachuck *et al.*, 1977). According to Tkachuck *et al.* (1977) vitamin B_{12} is converted into a coenzyme (methylmalonyl-CoA mutase) which is used in the reverse pathway reaction of succinate to propionate. In the process, ATP is also formed as one of the products of this reaction. The authors conclude that organisms with high concentrations of vitamin B_{12}, which accumulate propionate rather than succinate as a product of anerobic energy metabolism, may have the advantage of an increased energy yield (ATP) from their substrates. Apparently, all known helminths that form propionate as a major product of fermentation, rather than primarily succinate, lactate or products unrelated to propionate, contain high concentrations of vitamin B_{12}; species without vitamin B_{12} do not form propionate. It may be inferred from these observations that caryophyllids form propionate. However, nothing is known of the major metabolic end products of any caryophyllid, although there is some information on the enzymes of the glycolytic sequence and the tricarboxylic acid cycle in *Khawia sinensis* (Körting, 1976). If propionate is an end product in the energy metabolism of caryophyllids, accumulations of vitamin B_{12} could function to increase the energy *available for egg production*. If applied to egg production, the *increase* in egg production would help compensate for a lowered reproductive potential resulting from the monozoic body plan and the lack of asexual reproduction. Any physiological process that would increase fecundity would, of course, have adaptive and selective value.

H.   C-TYPE VIRUS PARTICLES

First described from the excretory tubules of a pseudophyllid *Sparganum proliferum*, by Mueller and Strano (1974) possible C-type virus particles have subsequently been described from the following species and stages in the same order: *Spirometra mansonoides*, coracidia, procercoid, plerocercoid and adult; *Diphyllobothrium ditremum*, *D. sebago* and *Ligula intestinalis*, plerocercoids (Dougherty *et al.*, 1975). Different, nodular-like particles, some branched but not showing internal structures were also found lining the excretory ducts of an unknown proteocephalid plerocercoid, and adults of cyclophyllids *Taenia taeniaformis*, *Dipylidium caninum* and *Hymenolepis diminuta*. Earlier workers, beginning with Race *et al.* (1966), had also found nodular-like projections or processes in the excretory canals of various other cyclophyllids. Although there is some doubt that the particles from *S. mansonoides* are in fact viruses, because of the apparent absence of nucleic acids, the presence of entities lining the excretory ducts of such diverse cestodes as cyclophyllids and pseudophyllids suggests that they may be characteristic of the excretory systems of all cestodes. Recent studies by Edwards and Mueller (1978) clearly indicate that this supposition is not true because they found that C-type virus particles were not present in the following nine caryophyllid cestodes from three families: Caryophyllaeidae, *Glaridacris laruei*, *G. terebrans*, *Monobothrium ulmeri*, *Isoglaridacris folius*, *I. calentinei* and *Hunterella nodulosa;* Lytocestidae, *Atractolytocestus huronensis* and *Caryophyllaeides fennica;* Capingentidae, *Capingens singularis*. If these data are true, then one may wonder if the absence of these structures, whether viruses or simple modified microvilli, from the excretory ducts of caryophyllids is evidence of functional differences between the excretory systems of these cestodes and others or whether they reflect *a basic difference between the excretory system of monozoic and polyzoic tapeworms*. Clearly, more specific information on the origin, development and function of these C-type virus particles, or microvilli, is needed before their evolutionary significance and possible selective value can be assessed.

IV. *Archigetes* EVOLUTION

A. GENERAL

As the only cestode that has a complete, one-host life cycle in an invertebrate (see Section IC1, *Cyathocephalus*), *Archigetes* occupies a singular position in any discussion of cestode evolution. As its name indicates, *Archigetes* has been regarded either as a primitive ancient tapeworm, and for some as an ancestral cestode (Baer, 1952), modified form with a precestode cycle (Freeman, 1973; Demshin, 1971), relict of the most primitive one-host cycle (Jarecka, 1975), or an example of a cercomeromorphaean progenitor (Malmberg, 1974). Despite its long history, there are still questions regarding the evolutionary significance of *Archigetes* and there is little agreement whether or not it can be regarded as a procestode, an adult cestode, or as evidence that cestodes were originally parasites of invertebrates.

Before any of these questions can be answered, the status of *Archigetes* with respect to other caryophyllids should be clarified. The systematics and biology of *Archigetes* have been reviewed by Kennedy (1965a, b), Calentine (1962) and Kulakovskaya (1962b) and need not be repeated in detail here. It is quite possible, judging from the extensive synonomy of the genus that includes *Brachyurus*, *Paraglaridacris*, and in part *Glaridacris* and *Biacetabulum*, that *Archigetes* is in fact a collection of species from various genera. For the present, however, it is the biology and development of *Archigetes* and not the systematics that is relevant here. Except for *A. cryptobothrius*, found only in tubificids, the other five species apparently mature in both tubificids and fish. Since the stage in fish is no different from any other caryophyllid, it is the stage in the tubificid that concerns us here.

B. EGG DISPERSAL

Stages of *Archigetes* in the coelom of tubificids have two basic and unmistakable characteristics of immature (procercoid-like) stages: a cercomer and a non-functional genital pore. Both of these characteristics are absent from stages found in fish. The cercomer, a common feature on the immature stages of cestodes of various orders (Freeman, 1973), is indeed a recognized and genuine feature of "larval" cestodes and needs no elaboration. The genital pore, the second characteristic, is rendered non-functional by a covering cuticular layer (Calentine, 1964). As eggs are produced they may either accumulate in a greatly distended uterus that may occupy most of the whole worm (Fig. 7b) or they may fill a cuticular pouch formed by a splitting of the outer and inner cuticular layer (Figs 7a, 7ci and 7cii), according to Calentine (1964). Sometimes both types of egg retention take place in the same species (i.e., *A. sieboldi*). Regardless of which type occurs, eggs can be liberated only by rupturing the cuticular layers of the worm itself, processes which kill the cestode. Such mechanisms of egg deposition are surely abnormal, even aberrant. The chief way that cestode eggs can leave the infected tubificid is by rupture of the body wall and release of the cestode into the substrate where it subsequently dies, liberating the eggs. Effective egg dissemination is thus accomplished by the death of both host and parasite, clearly

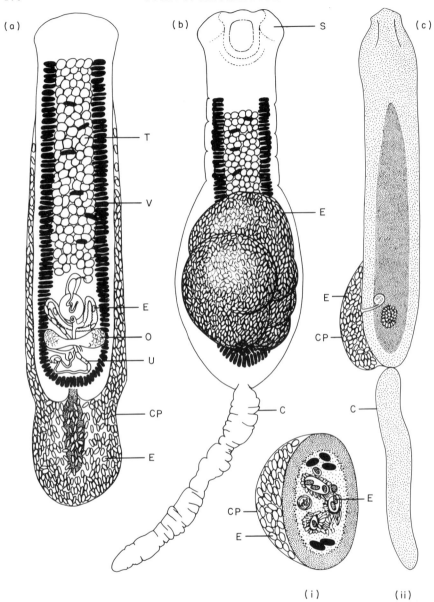

FIG. 7. (a) *Archigetes cryptobothrius* with cuticular pouch (CP) filled with eggs (E). Testes (T), vitellaria (V), ovary (O), and uterus (U) are clearly visible. (Adapted from Wiśniewski, 1930). (b) *Archigetes sieboldi* with greatly swollen uterus filled with eggs (E); well developed scolex (S) and cercomer (C) are evident. (Adapted from Wiśniewski, 1930.) (c) *Archigetes sieboldi*. (i) Cross-section through posterior portion of cercomer bearing individual illustrating cuticular pouch (CP) filled with eggs (E), also visible in the uterus. (ii) Lateral view with cuticular pouch (CP) distended with eggs (E) but not involving the cercomer (C). (Adapted from Calentine and DeLong, 1966.)

an abnormal, perhaps aberrant, mode of completing the reproductive cycle. In view of this mode of egg dissemination, the presence of a cercomer, the coelom or seminal vesicle habitat, and normally reproducing adults in fish, the conclusion is inescapable that the egg-producing stage in the tubificid is a progenetic cestode, secondarily derived from an intestinal stage in fish. Some species such as *A. limnodrili* and *A. sieboldi* appear to be able to complete and sustain a cycle in tubificids (Kennedy, 1965b; Calentine and DeLong, 1966, Nybelin, 1962), others such as *A. iowensis* can occasionally mature in tubificids but are more common in fish (Calentine, 1964); strain differences may explain why there have been conflicting data on whether a fish can be in the cycle. Additional work is necessary to determine if any species of *Archigetes* is facultatively progenetic.

C. PROGENESIS

Comparing the development of progenetic *Archigetes* with that of other caryophyllids, one finds one basic difference between the two: maturation in *Archigetes* occurs before the cercomer is lost, in the other, after. There is one immediate consequence of this acceleration: maturation is in the coelom of the tubificid, in other caryophyllids it is in the intestine of the vertebrate host. The developmental difference between the coelom stage and intestinal one can be very slight. The intestinal stage of *Archigetes* is identical to the coelom stage except that it lacks a cercomer and has a functional genital pore. Furthermore, there is a pronounced tendency toward progenesis in many caryophyllids, for example, in such genera as *Caryophyllaeus* (Sekutowicz, 1934), Kulakovskaya, 1962a; Mackiewicz, 1972; *Hunterella, Monobothrium, Glaridacris, Biacetabulum* (Calentine, 1965, 1967) and *Khawia* (Demshin, 1978) reproductive organs reach various degrees of development (but not maturity) in the cercomer-bearing stage in the tubificid coelom. We therefore have a continuum with *Archigetes* at one end and various other examples approaching complete progenesis. But what factors may have selected for complete progenesis in *Archigetes*?

A possible answer may be related to small size. If small size is the primary object of selection for progenesis, as Gould (1977) suggested for a number of free-living organisms where space is a limiting factor, then progenesis may be a passive consequence and not itself the product of selection. Space limitations within the tubificid coelom or seminal vesicle may exert a selective influence for small size. If space (and therefore some resources) is one of the limiting factors, it is therefore not surprising that progenesis is most common in small forms, such as *Archigetes*, that are generally less than 3 mm long. Should this be the case, then on theoretical grounds one can expect progenetic stages in other small forms as *Penarchigetes* (1·7 mm) and *Balanotaenia newguinensis* (0·85 mm). In my view, it is quite possible that progenesis may not be restricted to *Archigetes*, a fact also suggested by the *Biacetabulum–Archigetes* controversy (see Calentine, 1964, for review) as well as synonymies with other genera (see above).

According to Gould (1977), who accepts the classical view of *Archigetes*, progenesis in *Archigetes* has been ecologically determined by parasitism. Just how parasitism acts as a selective mechanism, however, is not explained. The question remains: what are the biological or ecological factors that select for progenesis whose consequence is a simplification of the life cycle? Among the possibilities, is that progenesis leads to increased fecundity, a principal attribute of parasites. Unfortunately, we do not have any data on comparative egg production for progenetic *Archigetes* and those in the two-host cycle. We know that development to the gravid stage in tubificids is very slow, 140 days for *A. limnodrili* (Kennedy, 1965a), 120 for *A. sieboldi* (Calentine and DeLong, 1966) and 100 days for *A. iowensis* (Calentine, 1964). Presumably development would be faster in the nutrient-rich intestine of a fish. If the shortest two-host cycle is less than the shortest for a one-host cycle, then there would appear to be no selective advantage for the progenetic cycle in terms of fecundity. On the other hand, it would appear that a shorter one-host cycle could produce more generations with a resultant net gain in fecundity. As Lewontin (1965) and MacArthur and Wilson (1967) have concluded, a speeding up of maturity can be more effective in the long run than increasing the fecundity or fertility per individual. These conclusions, however, are based on organisms that prolong the reproductive period by starting earlier, of obvious selective value in colonizers. In *Archigetes*, however, eggs are shed all at once because the genital pore is *non-functional* and eggs are retained until the cestode ruptures the body wall of the tubificid and dies outside the host. Unless progenesis so shortens the generation time in a one-host cycle to allow for more generations than the two-host cycle (3 to 4 months can hardly be considered short), it is difficult to see how selection is for fecundity. Even if the generation time was shortened, oligochaetes seldom have more than six parasites and thus the overall increase in egg production is not large —certainly not compared with worms in fish where there may be over 1000 individuals (Calentine, 1964). It would appear that progenesis does not greatly increase fecundity over that from a population in fish; there must, therefore, be some other basis for selection.

D. SELECTION

That basis may be related to the fact that on theoretical grounds (no actual information is available) fewer eggs are probably needed to sustain a population of *Archigetes* in a population of tubificids than in a fish population. Liberated eggs from a decaying cestode in mud can be ingested readily by other tubificids, an event of high probability because of the clumping habits of some tubificids (Brinkhurst and Jamieson, 1971) or the general abundance of oligochaetes in the benthos. Selection would favour populations of progenetic *Archigetes* because there probably would be a greater probability of their eggs being eaten by another tubificid over an egg expelled from an intestinal dwelling (fish) *Archigetes*. In my view, progenesis may so increase the probability of infection by another tubificid that the genome for this precocious development would be selected for. In this way, selection would be for limited dispersal in contrast to the wide dispersal found in the two-host cycle.

Depending on the degree of isolation and selection pressure for the progenetic trait, it is possible to get a series of *Archigetes* species showing various degrees of selection for one- and two-host cycles. With *A. iowensis*, Calentine's (1964) data suggested that only eggs from progenetic individuals would produce other progenetic cestodes; only 3·2% of 567 naturally infected tubificids were progenetic, yet 80% of the worms in carp had eggs. Clearly fish maintained the tubificid infection in nature but there was evidently a small population of cestodes with a one-host cycle. It would appear that we have here a one-host (progenetic) and two-host (non-progenetic) cycle in the same species. In *A. sieboldi*, on the other hand, the fish host appears to play a minor role in maintaining the cycle because Calentine and DeLong (1966) found that 56% of 472 procercoids were gravid in *Limnodrilus* whereas 19 *Cyprinus carpio* harboured only five gravid cestodes. Nybelin (1962) considered fish to be accidental hosts of *A. sieboldi*. In *A. limnodrili*, at the end of the series, it appears that only progenetic development takes place (Kennedy, 1965a). Its patchy distribution in *Limnodrilus*, a result of limited egg dispersal, is consistent with the hypothesis that fish are not necessary where a progenetic population (or species) has evolved. Progenesis, then, favours the one-host cycle, not because it increases fecundity, but more likely because it increases the probability of tubificid to tubificid infection.

How does one now relate the one-host and two-host cycles to the evolution and biology of *Archigetes*? Clearly, both cycles have responded to different selective pressures and in so doing have greatly increased the probability of survival for the species. The two-host cycle, normal for caryophyllids and in my view the original cycle, has a fish as the primary agent of dispersal. The risk of widely dispersed eggs not being ingested by tubificids is partially compensated for by the increased egg production from the population of intestinal cestodes; for example Calentine (1964) found up to 1523 *A. iowensis* in one carp. Success of such a cycle depends on many factors, among them the number of fish; if high, the two-host cycle can be completed more easily than when there are few fish. It is when the fish population is low, or fish are absent, that the one-host cycle has a greater selective value. Under conditions of low oxygen or high water temperatures, or any other that would limit or kill off the fish population, *Archigetes* would be able to survive without the definitive host. Should the definitive host (vertebrate) *never* return the species is still able to survive. Perhaps some species (*A. sieboldi*, *A. limnodrili* and others) have become facultative progenetic cestodes with alternate cycles in response to adverse environmental conditions or fluctuating fish populations. By being able to switch from one cycle to another, or have the potential to do so, *Archigetes* is able to survive under a greater range of stress conditions than any other caryophyllid. In terms of r–K strategies, it would appear that the two-host cycle is toward r-selection, whereas the one-host cycle is toward the K side of the r–K continuum. Whatever the genetic mechanism, production of two types of eggs or development of physiological strains, we have in *Archigetes* a remarkable cestode, one that survives by a mixture of one- and two-host cycles, rather than a cestode by-passed by evolution and surviving as an ancestral form or relict of some prehistoric tapeworm.

E. CONCLUSIONS

In the light of the developmental status of a progenetic organism and the role of natural selection in determining ecological and host–parasite relationships, what are the answers to some basic questions regarding the evolutionary position of *Archigetes* in the one-host cycle?

1. *Is Archigetes an adult cestode?* If by "adult" is meant a final reproducing stage possessing only adult morphological characteristics then the stage in the intestine of fish is an adult. The gravid stage in the tubificid is not an adult in the same sense because of the presence of *bona fide* "larval" characteristics, i.e., cercomer and non-functional genital pores. An organism cannot be both progenetic and an adult. Of course, if one regards any reproducing stage of a caryophyllid as an adult, regardless of morphological features, then *Archigetes* is an adult.

2. *Is Archigetes an ancestral cestode?* If one is convinced that *Archigetes* is an adult stage that has evolved as a parasite of tubificids, its original and only host, and is now evolving or has evolved a two-host cycle by adding a vertebrate host—the answer is yes. In my view, the evidence that (a) *Archigetes* is secondarily derived from a vertebrate dwelling stage, and (b) tubificids are intermediate hosts, is persuasive if not conclusive. Accordingly, the *Archigetes* one-host cycle must have evolved after (not before, and possibly concurrent with) the two-host cycle; the one-host cycle is thus a simplificaiton of a pre-existing two-host cycle. The one-host cycle can be adequately derived from a two-host cycle by the application of ecological theory, an elaboration of the dynamics of natural selection. Deriving a two-host cycle *from* an ancestral one-host cycle, on the other hand, requires numerous assumptions and a chain of events that become difficult to justify using ecological theory and natural selection. Viewed in this perspective, *Archigetes* is not an ancestral cestode (or precestode, procestode, relict cestode, Urcestode, or cercomeromorphean progenitor) but a form that has diverged from a two-host cycle stock through selection of progenesis as an ecological strategy. Just as progenetic digenea or monogenea are not regarded as ancestral forms, nor should *Archigetes* be.

3. *Is Archigetes more primitive than other caryophyllids?* No, if by primitive one means original, earliest, primary, ancestral or ancient; a secondarily derived form cannot be, by definition, more primitive than its ancestors. Yes, if one can provide proof that *Archigetes* is ancestral to other forms.

4. *Was the original host of Archigetes an invertebrate or vertebrate?* The answer to this question, pondered for centuries, is the same for *Archigetes* as it is for polyzoic cestodes. It is beyond the scope of this paper to review all the theories on the first hosts of cestodes.

Rosen (1918) felt that *Archigetes* is a species "*sui generis*" and is a primitive cestode of invertebrates whose relationship to other cestodes is expressed in the statement: *Fecampia–Archigetes; Archigetes–Procercoide.* He clearly felt that the parental habit of *Archigetes* was not evidence of regression in the cycle, citing the symbiotic rhabdocoel, *Fecampia* as an example of another flatworm with similar habits. Furthermore, he regarded the well-developed scolex, as he did the oncosphere, as an adaptation for parasitism and not proof that *Archigetes* was once an intestinal parasite. These views have proved to be the strongest in favour of the invertebrate origin of cestodes through an *Archigetes*-like ancestor. However, in the light of subsequent information on the morphology (i.e., cuticular pouch) and life cycle, these views should be critically reevaluated. In my opinion, it is not warranted to use *Archigetes* to support the thesis that invertebrates were the original hosts because there is no conclusive proof that the vertebrate stage followed the invertebrate one. On the contrary, the opposite appears to be true. If invertebrates were the original hosts, even in some cases, why are there no examples living today? To say they are extinct begs the question, as does indicating that evolution selected for the more complex two-host cycle.

V. EVOLUTIONARY SCHEMES (1971–1979)

A.   GENERAL: ENTODERM QUESTION

There has been a long history of speculation regarding the role played by caryophyllids in cestode evolution; much of this history was reviewed by Mackiewicz (1972). Hypotheses concerning the primary or secondary nature of the monozoic condition and whether or not invertebrates or vertebrates were the original hosts form the basis of most of the speculation. With few exceptions (Cameron, 1956, 1964, for example) past hypotheses have assumed that cestodes arose through a rhabdocoel line allied to the Dalyellioida turbellarians and through the parasitic habit gradually lost all traces of a digestive system. These assumptions, together with additional arguments linking the cestoda with the monogenea (Bychowsky, 1957; Llewellyn, 1965) have formed the foundations of perhaps the most widely known recent hypothesis of cestode origin. As some of the basic assumptions of this hypothesis have been challenged, new theories of cestode origins and especially the role of caryophyllids in cestode evolution have developed.

1. *Did cestodes have a gut?*

One of the assumptions questioned recently is that cestodes have secondarily lost a gut. It has long been a zoological truth that cestodes have lost their gut through evolution in a nutrient-rich habitat, the intestine of vertebrates. There are no creditable studies, however, that present direct evidence that a gut or rudiments of one occur anywhere in the development of any cestode. By linking the cestodes with rhabdocoels or monogenea it is obvious that the gut *had* to be lost in order to make such a scheme reasonable. Inextricably tied to the question of a former gut is a related one dealing with basic cestode embryology: do cestodes have entoderm?

Both of these points have been challenged by Logacev (1968, 1970) whose influence is seen in some more recent treatments of cestode evolution. Summarizing his extensive studies on the embryology and development of cestodes, Logacev (1970) concludes that cestode cleavage is irregular and leads to a syncytial mass in which there is no entodermal layer; nowhere in cestode ontogeny is there any structure that is even remotely homologous to an intestine. Citing also the macrophagic mode of feeding through the tegument as the most primitive mode of feeding, Logacev (1970) concludes that cestodes are a phylogenetically independent branch of the intestineless Acoela. He cites the work of Severtov (1945) who concluded, with respect to the nutritional function of the tegument of endoparasites, that "the primary process depends on formation of new adaptations and not on reduction". According to Logacev, the lack of intestine, mode of feeding, absence of basal membrane and histological differentiation of inner and outer tissue link the Acoela and Cestoda together. The coracidium of the Pseudophyllidea is said to recapitulate the early turbellarian ancestor and the non-cellular character of the cestode embryo is considered a recapitulation of the phagocytoblastic or syncytial digestive parenchyma of the primitive intestineless turbellaria. His paper contains additional information on cestode histogenesis, strobilization, histology of proglottid separation, embryology, ecology and genetics which is brought to bear on his main thesis—that cestodes descended from a two-layered ancestor with no gut and are a dead-end branch evolved from ancient acoelous turbellarians.

It is well to remember that Logacev's hypothesis uses a different frame of reference, not that of classical germ-layer theory with the well-known ectoderm and mesoderm concepts, but that of the primary two-layer theory with kinetoblast and phagocytoblast. He is therefore quite correct in stating cestodes have no entodermal derivatives since the concept of entoderm is not incorporated in his basic theory of embryology for primitive metazoans. In the lower metazoa, which include the acoelous turbellarians, the kinetoblast forms the external layer and the phagocytoblast, the internal structures. Since the boundaries between these two layers are not always distinct, organs may have tissues partly derived from each layer. How the kinetoblast–phagocytoblast hypothesis applies to metazoan evolution is discussed at great length by Beklemishev (1969, Vol. 2; 194) who says of the Cestoda that there is a "complete reduction of the central phagocytoblast and of the entire internal digestive apparatus, as a result of their parasitism in the gut of vertebrates". He concludes that the absorptive function passes to the kinetoblast which forms the external epithelium (tegument) and that the low organization level of lower worms is why cestodes and acanthocephala lose the gut. Unlike Logacev, Beklemishev, using the same frame of reference, derives cestodes from the Dalyellioida rhabdocoels through reduction and loss of the gut.

Whether or not cestodes have entoderm may thus be more a question of definition rather than actual fact; however, since the term has been used so widely in the literature it will be retained here. According to Schauinsland (1885) and Vergeer (1936), who have done the most complete studies on pseudophyllidean embryology, the ciliated embryophore of *D. latum* is ectoderm and the inner part, mesoderm; Hyman (1951) designates the inner mass "mesentoderm". As pointed out by Burt (1963), such a view of an entodermally derived tegument would be consistent from a functional and morphological point of view. Cameron (1956) has even suggested that the adult cestode may be entirely entodermal, hence explaining the lack of a gut anywhere in cestode development; on the other hand, he also indicates that cestodes may be an entodermal sac whose lumen has been filled with mesoderm. Ogren (1956, p. 423), the foremost student of cestode embryology in the western hemisphere, concludes that "identification of germ layers is not possible with certainty at this time", after studying the embryology of *Mesocestoides corti*. He further states that the body covering of *M. corti* is not a digestive epithelium and therefore is not entodermal; there are no entodermal derivatives in *M. corti* according to Ogren (1956), who concludes (p. 423) "In the present state of our knowledge of tapeworm embryology, all that can be said is that the tapeworm strobila is predominantly 'mesodermal'". The concept of entoderm or any other germ layer is not used in the recent studies of cestode embryology by Euzet and Mokhtar-Maamouri (1976). It is perhaps significant that Rybicka (1966) does not use any germ-layer concept in her extensive review of embryogenesis in cestodes. When one tries in vain to reconcile cestode embryology with classical germ-layer theory it becomes clear why DeBeer (1958) considered the germ-layer theory of Haeckel as "fallacious".

From the above it is apparent that there is little question that cestodes lack a gut anywhere in their development, but how does the classical germ-layer theory relate to cestode embryology? Some attempt to resolve the issue has been made by Bazitov (1974) who studied the embryology of *Hymenolepis nana* in serial sections (5 μm, thick) prepared with a variety of stains including methyl green pyronine. Briefly, he found that a blastocoel is formed and that the two pale entodermal blastomeres protruding into the blastocoel gradually became lysed leaving an embryo with ectoderm and ectomesoderm, the latter eventually filling the blastocoel. Morphogenetic migrations and distinct layers are not evident because of the low level of complexity of the onchosphere; a two-layered gastrula, as generally understood, is not formed. In the opinion of Bazitov (1974), only the germ-layer theory can give a satisfactory explanation of the embryology of *H. nana*, even though there are some abnormalities in development such as the abortive character of gastrulation. In a subsequent paper on the embryology of the cyclophyllidean *Microsomacanthus paramicrosoma*, Bazitov and Lapkalo (1977) strengthen the evidence that entodermal blastomeres (micromeres), homologous to entomesodermal micromeres of other flatworms, abort at the coeloblastula stage (Figs. 8a–c). They report that in gastrulation of cestodes the vitelline-rich macromeres migrate to the surface of the embryo, whereas in the lower turbellaria they become part of the inner cell mass of entomesoderm. The macromeres form the vitelline envelope at the surface and function in the utilization of vitelline material and exogenous food. According to Bazitov and Lapkalo (1977, p. 110), "In connection with this the necessity for the formation of a gut disappeared. It is necessary to think that the reduction of the alimentary system in cestodes is tied not only to the changes in perspective significance of macromeres but also to the degeneration of

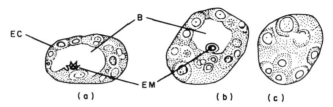

Fɪɢ. 8. Sections of embryos of *Microsomacanthus paramicrosoma* (a) Gastrula with blastocoel (B), ectoderm (EC), immigrated entomesodermal micromere (EM). (b) and (c) Con-
secutive serial sections at stage when blastocoel is being filled up. (Adapted from Bazitov
and Lapkalo, 1977.)

micromeres that had immigrated". Except for terminology, these views are
somewhat similar to those of Logachev (1970) regarding the trophic functions
of the outer layer. They also conclude that the egg has features of duet and
spiral fission characteristic of the lower groups of turbellaria, in contrast to
Costello and Henley (1976) who do not mention cestodes but acknowledge
that acoel turbellarians have a modified spiral cleavage.

Related to the entodermal question in cestodes, is a similar one in acoel
turbellarians, the theoretical ancestral stock of cestodes according to Logacev
(1970). The literature on this question is quite extensive and will not be
reviewed here. As an example of how views differ, Boyer (1971) indicates
that the endoderm and mesoderm of the Acoela are not clearly defined
because of the absence of the 4d cell, yet Jennings (1974, p. 175) regards the
intestine in Acoela as "... a syncytial undifferentiated mass of endoderm,
containing numerous vacuoles in which digestion occurs". Costello and
Henley (1976) appear to follow earlier workers and designate the non-
ectodermal part of the early embryo as "internal parenchymal mass"; the
question of entoderm or entomesoderm is not dealt with. From the above
discussion, it is evident that there are no clear answers regarding the presence
of entoderm in cestodes. The evidence appears in favour of some vestige of
entodermal (or entomesodermal) cells which abort in early development with
a change toward a trophic function for the outer layer of the embryo. Whether
this interpretation can be used as evidence of a reduction of a gut remains
to be seen as additional facts on cestode embryology emerge. On the basis of
these studies, it appears to me that cestodes did not lose a gut by reduction
but never had one, or if lost it was correlated with the parasitic habit and not
necessarily an intestinal one. It is possible, however, with current data to
make arguments for either an acoel ancestor or, as more generally believed,
a rhabdocoel. A review of some of the more recent schemes of cestode and
caryophyllid evolution will illustrate which arguments have had the greatest
influence. So diverse are the views that the papers are taken in chronological
order.

B. 1971–1975

Demshin (1971) is among the first to accept Logachev's (1970) views. The
monogenea are not regarded as being in the cestode line of evolution for a

number of reasons, among them characteristics of life cycle, morphology, ecology and, most important, a lack of explanation for one branch (cestodes) losing a germ layer and another (monogeans) keeping it. As a consequence, he believes that cestodes are an independent branch that evolved from an acoelous turbellarian ancestor and first became parasitic in the *body cavity of oligochaetes*. Oligochaetes were suitable hosts primarily because of their large size and also because of their population density, wide dispersal, and benthic habit, the latter allowing a greater exposure to eggs than by other invertebrates. Endoparasitic forms of acoels and rhabdocoels evolved in molluscs, starfish, sea urchins and decapod crustaceans, whereas *Archigetes* evolved in oligochaetes. Demshin (1971) thus regards *Archigetes* as originally parasitic in oligochaetes and unchanged until the present. As vertebrates evolved, the one-host cycle gradually developed into a two-host one with the obligate oligochaete phase as a permanent part of the cycle. *Archigetes* is thus considered a young cestode, basically organized as an adult but differing in size, state of development, and having fewer adaptive features. According to Demshin the procercoid is a young stage (juvenile?) and not a larval one because it does not have an embryonic character to its organization. He does not consider the coelom habit as a larval characteristic; nothing is said about the significance of the cercomer or of a non-functional genital pore. This latter characteristic of *Archigetes* is apparently overlooked, for he states that the cestode emerges from the ruptured tubificid and *deposits* eggs before perishing. Such a sequence overlooks the fact that *Archigetes* cannot deposit eggs at all but must rupture the tegument or cuticular pouch, either process killing the worm.

With the appearance of vertebrates that fed on oligochaetes, some *Archigetes*-like cestodes matured in the fish intestine and radiated into new forms, evolving cycles with oligochaetes as intermediate hosts. Other cestodes, however, did not mature in the gut but penetrated into the *body cavity* (how is not made clear by Demshin) where they matured. With the death of the vertebrate host the cestode eggs were liberated and became accessible to other invertebrates such as zooplankton. A series of adaptations, such as cilia on the oncosphere, coevolved with the development of a crustacean–fish cycle. In some cases, such as *Ligula*, strobilization began in the plerocercoid stage. Demshin (1971) further reasons that as in the cycle of *Ligula* the large plerocercoid rendered an infected fish more susceptible to predation thus adding a third, but warm-blooded host, to the cycle. In the final vertebrate, the strobilate stage completed development. In other cases, plerocercoids were liberated in the gut of predacious fish and matured to give strobilate forms. According to Demshin, the Pseudophyllidea were the first strobilate cestodes to evolve; with the rise of terrestrial vertebrates the cestodes radiated into different hosts and along different lines. His evolutionary scenario is thus: acoela ancestor to caryophyllid-oligochaete cycle (*Archigetes*) from which one branch evolves no further but the other incorporates zooplankton and eventually evolves to strobilate forms.

On the whole, this scheme is appealing because it starts with a supposedly primitive form (*Archigetes*) and derives the more complex cestodes. Unfortunately, he has completely overlooked the possibility of progenesis playing a part in caryophyllid evolution or even with *Ligula*, which has advanced development while still in the body cavity of fish. Furthermore, he does not deal effectively with the question of why oligochaetes are not more fully exploited by other cestodes since they were the original hosts of tapeworms. These and other important questions are dealt with in the paper by Mamaev (1975).

Czaplinski (1972) also accepts Logacev's (1970) basic thesis that cestodes lack entoderm. From a planula-like ancestor (acoelous prototurbellarian) he has two lines of evolution: one with entoderm that gives rise to a rhabdocoel ancestor from which the Trematoda, Rhabdocoela, Temnocephala, Unonelloidea and Monogenea arise and a second line, without entoderm, that gives rise to Acoela, Gyrocotyloidea, Cestodaria and Cestoda. The difossate cestodes diverge from the cestodaria stem but the tetrafossate ones branch off the ancestral line for all groups. Cestodes are said to have first parasitized invertebrates and only radiated further with the appearance of vertebrates. His scheme makes no mention of the Caryophyllidea, unfortunately.

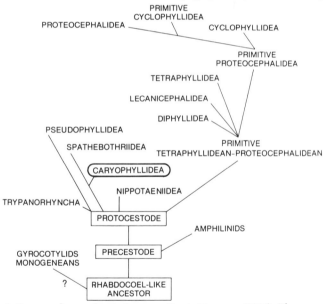

FIG. 9. Evolutionary scheme of cestodes according to Freeman (1973). The precestode stage had an adult that was single-suckered and free-living; the onchospere came from an anoperculate egg and developed parenterally in an invertebrate; the plerocercoid-preadult was tailed and free-swimming. The protocestode stage had an adult that was single-suckered and in the vertebrate gut; the oncosphere came from an anoperculate egg; and the plerocercoid had a metamorphosis (?) parenterally in an invertebrate. Nothing should be implied from the angle or length of connecting lines. (Adapted and simplified from Freeman, 1973.)

By far the most extensive, often brilliant, treatment of cestode evolution in recent times is that of Freeman (1973). Although he does not bring up the question of entoderm in cestodes or mention the possibility of an acoel ancestor, his arguments are generally well-presented and with fresh insights. His evolutionary scheme, somewhat simplified, is presented in Fig. 9. Clearly he regards cestodes as a polyphyletic group away from the monogenean line with caryophyllideans as a dead-end branch on the same line as the Spathebothriidea; the paper must be read to get the detailed rationale for such a scheme. Of particular interest for caryophyllid evolution, is the incorporation of a hypothetical precestode (in invertebrates) and protocestode (in vertebrates) stage between the rhabdocoel-like free-living ancestor and cestodes. It is the precestode cycle, consisting of a free-living, tailed pre-adult or adult that concerns us here. The fact that *Archigetes* is the only cestode that matures in a parenteral site suggested to Freeman (1973) that the precestode adult stage was free-living and the preadult stage was parasitic in invertebrates. A similar view is expressed by Stunkard (1975).

According to Freeman (1973), *Archigetes* is, as Rosen (1918) believed, a primitive adult cestode, or "caudate adult" in his new terminology. His interpretation of *Archigetes* is quite unlike that presented anywhere else. For example, he believes that *Archigetes* has a free-living stage, though brief, after it ruptures free of the tubificid and because of this is an example of the primitive precestode, the cercomer being a vestige of a former swimming organ. As far as I am aware, this is the first time that anyone has considered any stage of caryophyllids as free-living.

Unless one considers the brief period before death of a cestode expelled or ruptured free as free-living, there is no justification whatsoever for considering *Archigetes*, an obligate parasite, as having a free-living stage. Once free of the tubificid, *Archigetes* is doomed to decay; by no standard biological criterion can this period of slow death be called free-living. Unfortunately, other more important aspects of *Archigetes* morphology or biology are omitted. For example, there is no mention of the origin of the well-developed scolex, which is unlike the hypothetical single terminal sucker-like structure (unifossate) characteristic of precestodes (see Fig. 11a), or why *Archigetes* matures in a parenteral site, acknowledged to be the habitat of metacestodes. By failing to realize the significance of a non-functional genital pore and stretching the definition of free-living, Freeman (1973) has ascribed to *Archigetes* a far greater evolutionary role than is warranted; it has also prevented a consideration of progenesis as a possible explanation for the precocious maturity in the invertebrate host.

As for the evolution of other caryophyllideans, Freeman (1973) believes that they were primarily monozoic and evolved from a unifossate, monozoic ancestor in the protocestode stage. Little is said of the role of tubificids. The concept of progenesis receives little attention.

Without reference to Logachev (1970), Demshin (1971) or Freeman (1973), Malmberg (1974) develops a line of evolution for the Cercomeromorphae Bychowsky, 1937 (including the Caryophyllidea), utilizing the protonephridial system to show relationships. Some of the key points of his scheme that differ from an earlier one (Malmberg, 1971) are (a) the ancestor was a parasitic "acoelomatic creature", whose eggs were eaten by mud eaters, for example, gastropods and annelids (oligochaetes), (b) cestodes never had an intestine in any stage of their ontogeny, (c) a cercomer evolved *after* forms became parasitic, (d) caryophyllideans or monozoic cestodes are genuine non-progenetic or non-neotenic adults, and (e) strobilization evolved along a separate line with *intestinal* parasitism. Like Demshin (1971), Malmberg derives the cestodes from a caryophyllidean line through oligochaetes. Of the cercomeromorphean line he says (Malmberg, 1974, p. 77), "The body shape and the type of endoparasitism of these cercomeromorphean progenitors are best preserved in adult caryophyllideans with a retained cercomer, *i.e., Archigetes*". According to Malmberg, the incorporation of a vertebrate into the cycle allowed the procercoid stage to become a plerocercoid in a parenteral site; with the addition of a final predator, these plerocercoids developed a strobila and scolex under the influence of the intestinal habitat in the predator. This part of the hypothesis is very much like that of Freeman (1973) and others; however, there is no explanation why there is a well-developed scolex in the parenteral stages of caryophyllids, before the intestinal phase is reached. Malmberg believes that the caryophyllids evolved before other cestodes and that the monozoic and polyzoic body plans evolved along separate lines.

The preceding hypotheses of Demshin (1971) and Malmberg (1974) are discussed in detail by Mamaev (1975) who strongly disagrees with the basic thesis that *Archigetes* is a protocestode and that oligochaetes were the initial hosts of cestodes. Citing the non-functional genital pore, and references to *Archigetes* as a classical example of a progenetic procercoid (see Mackiewicz, 1972 for review), he rejects the view that *Archigetes* is an adult cestode and thus argues that it cannot be used as the stem form giving rise to other cestodes. On the other hand, he agrees with Demshin that the cercomer *per se* is not evidence of a larval condition; instead, it is an ancestral remnant still present on the larval stage. He places considerable emphasis on the role of progenesis in determining the evolution of caryophyllids and links *Archigetes* with those having a two-host cycle. He rejects the notion that oligochaetes were the first hosts of cestodes by asking a number of searching questions. Why are there no monogeneans on oligochaetes? Why are there no adult cestodes in oligochaetes? Why are there no primitive cestodes in terrestrial oligochaetes? Why are the closest relatives of caryophyllideans—*Cyathocephalus* and *Diplocotyle*—in salmonids with stages in amphipods rather than oligochaetes? Lack of satisfactory answers leads Mamaev (1975) to conclude that the primary host of cestodes were freshwater fishes, not invertebrates.

Unlike any other contemporary author who has recently addressed the evolution of caryophyllids in detail, Mamaev (1975) places considerable emphasis on the role of progenesis. It is his view that progenesis in pseudophyllidean cestodes ought to lead to (a) shortening the life-span of the adult stage in definitive hosts, (b) retarding growth of strobila and perhaps making it shorter, (c) multiplication of genital complexes as in *Ligula*, and (d) disturbing the process of strobilization. A complete loss of strobilization resulting in a monozoic body form might also be a consequence of progenesis, according to Mamaev. He believes, and I concur, that progenesis is of frequent occurrence in the caryophyllids and pseudophyllids because of their primitive structure and the fact that larger intermediate hosts allowed for greater development of larval stages. He is of the opinion that one cannot dismiss the idea that the vertebrate phase of caryophyllids are progenetic plerocercoids but that there is an alternative view that they may be adults which have become secondarily monozoic under the influence of progenesis. A somewhat similar proposal was expressed earlier by Mackiewicz (1972). Despite the fact that he theorizes that loss of strobilization of the adults could lead to a monozoic body plan through progenesis, Mamaev feels that caryophyllideans are primarily monozoic.

Not long after Freeman's (1973) fine contribution, there appeared another quite similar one by Jarecka (1975) who concentrated on the evolution of cestode cycles. Her paper also deals with larval terminology to a great measure yet does not cite Freeman's (1973) key paper nor those of Logachev (1970) or Demshin (1971). Some of her conclusions regarding monozoic cestodes are (a) the first-stage larva is the oncosphere, the second develops in tissues or body cavity of the first host, and the third stage is sexually mature and may be monozoic or polyzoic, (b) of the four basic second stages (proceroid, cercoscolex, cysticercoid and cysticercus) the procercoid of monozoic cestodes is the most primitive and is parasitic in the body cavity of oligochaetes, the most primitive of aquatic coelomates infected with cestode larvae, (c) there is a correlation between the host evolutionary level and that of larval morphology, and (d) the scolex adhesive organs in the second stage (procercoid) are a result of "interstadial acceleration" or preadaptation to the conditions for the third stage. Jarecka considers that monozoic cestodes represent the most primitive group of cestodes; among her reasons are: morphology, "neoteny", and being parasitic in oligochaetes, the most primitive coelomates as hosts of gymnosomic cercoides (acystic larvae). Her reference to "neoteny" is left without explanation.

Her speculations on the origins of the cestode life cycle are unlike those of other workers. Jarecka speculates that oligochaetes were the original and only hosts that *swallowed* a primitive flatworm that had a cercomer. Within the intestine the hooks of the cercomer are used to penetrate through the gut wall into the coelom where the primitive parasites reached sexual maturity, dispersing their eggs after the death of the host. *Archigetes limnodrili* is thought to be a relic of such a one-host cycle. Primitive vertebrates ate the infected coelomates and gradually became definitive hosts. With the addition of the vertebrate to the protocestode life cycle the third stage evolved in the vertebrate gut. Despite this evolutionary scenario for the two-host cycle of monozoic forms (i.e., *Caryophyllaeus*), Jarecka (1975) is frank to admit (p. 110) "It is difficult at present to try the question of polyzoic tapeworms derived from the monozoic ones or evolved independently from an ancestor characterized by alteration of sexual and asexual generations". Her scheme adds very little to those before it and even adds complexity by not accounting for the fate of the oncosphere in the one-host cycle. Like Demshin (1971), she gives a prominent role to oligochaetes as the first hosts of cestodes.

c. 1976–1979

Kulakovskaya (1976), the foremost authority on caryophyllids in the eastern hemisphere, regards them as the most primitive group of cestodes, diverging from the dalyellid ancestors (Fig. 10) at the beginning of the Paleozoic or even in the Proterozoic and becoming parasitic in oligochaetes; other cestodes evolved after the appearance of arthropods. The fact that caryophyllid eggs lacked floating adaptations allowed them to sink and come into contact with the intermediate hosts, oligochaetes (see Section IIIC).

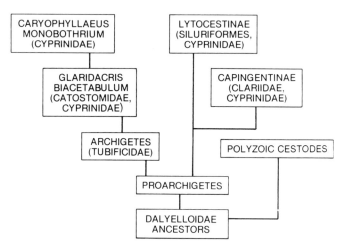

Fig. 10. A theoretical scheme of the phylogeny within the Caryophyllidea. The *Caryophyllaeus* branch and the subfamily designations used here correspond to the first three families on Table 3. (Adapted from Kulakovskaya, 1976.)

Because of their specificity for oligochaetes, Kulakovskaya believes that caryophyllids became fish parasites later than did other cestodes, certainly after cypriniformes had evolved, possibly in the middle Cretaceous. Another consequence of their long association with oligochaetes was that their evolution was retarded, thus explaining their monozoic form and low level of organization. Proof of their ancient mode of life, according to Kulakovskaya, is the maturation of some species in oligochaetes. Withing the group, the most primitive are the Caryophyllaeidae (Fig. 10) with medullary vitellaria, outer seminal vesicle and cercomer on sexually mature forms (i.e. *Archigetes*). Next are the Capingentidae with vitellaria partly cortical; and most advanced are the Lytocestidae, with cortical vitellaria, larger body size, inner seminal vesicle and "larval morphogenesis" in the intermediate host.

Of interest in this analysis by Kulakovskaya, is the reference to sinking eggs, a necessary prerequisite for tubificid infection. However, would one have not expected the eggs of the dalyellid ancestors (see Fig. 10) to have been already the sinking type since adults were free-living on the substratum? Her analysis places the caryophyllids as a polyphyletic group distinct from and along a separate evolutionary line from polyzoic cestodes. The inclusion of a proarchigetes stage is unlike that of other authors and appears to split the caryophyllids into two major lines whose primary difference would be the placement of the vitellaria with respect to the longitudinal muscles and presence of cercomer in mature forms. It seems doubtful that the former morphological feature should be accorded so much importance since it has been used only as a family or subfamily character. The cercomer is so clearly a transient structure common to all caryophyllids and should not be used to separate groups. Her paper has some interesting points, but she does not cite Demshin (1971), Freeman (1973), Malmberg (1974) or Mamaev (1975) and therefore does not deal with the question of a possible acoel ancestor, free-living stage, relationship to the monogenea, or the possible role that progenesis may have played.

The last two papers by Bazitov (1976) and Kulakovskaya and Demshin (1978) address themselves specifically to the phylogenetic relationships of the Caryophyllidea, but from different perspectives. Both are detailed treatments with bibliographies of 43 and 63 references respectively. However, neither paper seems aware of the work of Wardle et al. (1974) who place the Caryophyllidea, along with five other groups in a new class Cotyloda. Only the more significant observations or conclusions are reviewed here.

On the basis of a comparative study of cleavage, oncosphere morphology and histology of the parenchyma and subcuticula of caryophyllids with Cyclophyllidea and Pseudophyllidea, Bazitov (1976, p. 1785) concludes that, "Caryophyllids are a group distinct from cestodes and should be placed either in a subclass or in an independent *class* Caryophyllidea." (italics mine). No worker in modern times has made such a bold proposal elevating caryophyllids to the same level as the classes Trematoda, Monogenea, Turbellaria and Cestoidea. A subclass designation would presumably be comparable to subclass Cestodaria, and Cestoda or Eucestoda under the class Cestoidea as outlined by Noble and Noble (1971). Either position would elevate the caryophyllids to the status of a major platyhelminth group, a far cry from proposals that they are simply a family in the order Pseudophyllidea (Hyman, 1951; see Mackiewicz, 1972 for review of past schemes). Bazitov's proposal vividly illustrates how complex is the evidence upon which are made conclusions of the evolutionary and systematic position of these cestodes. What is the evidence used by Bazitov?

Cleavage pattern is one line of evidence. Using Motomura's (1929) study of the embryological development of *Archigetes appendiculatus* (it is the only one done on any caryophyllid), Bazitov regards the fact of indistinguishable macro- and micromeres by the 6-cell stage and an unequal second division as being a fundamental difference from strobilate cestodes in which there are two classes of blastomeres—micromeres and macromeres by the 4-cell stage.

Furthermore, there is no migration of blastomeres to the surface of the embryo nor is a cellular membrane formed. At the 4-cell stage of *Archigetes*, however, 2 macromeres and 2 micromeres *are* clearly distinguishable (Motomura, 1929; his Figs. 16 and 17). It is true that by the 6-cell stage the blastomeres become indistinguishable; however, based on nuclear characteristics (micronucleus=micromere; macronucleus=macromere) it is evident that at least two types of blastomeres can be distinguished through the oncosphere stage. It would appear to me that the alleged difference cited by Bazitov is not a functional one because macromeres and micromeres are indeed present although the size differential is small. Yet Motomura says little of cell migration and earlier he indicated that there is no blastocoel (Motomura, 1928), but the micromeres make the "mantel" or outer layer. Since there is no embryophore in caryophyllids, one cannot compare their development with that of the various pseudophyllids studied by Schauinsland (1885). It would appear that the embryology of caryophyllids, based on only one study, is different from that of pseudophyllids or other cestodes but it seems premature to regard such differences, which appear to be ones of degree rather than absolute, great enough to create a new class of flatworms.

Size of hooks of the oncosphere and morphogenesis of later stages are second lines of evidence. Again citing Motomura (1928), Bazitov attributes special significance to the fact that median hooks of *A. appendiculatus* are 13 μm and lateral ones 8 μm (in *K. sinensis* they are 12 μm and 9 μm respectively). This disparity in size is not typical of segmented cestodes according to Bazitov. I agree. Yet it is difficult to determine if such differences are indicative of separate class status. The absence of two membranes, present in pseudophyllidea, and presence of glands under the medial hooks as in cyclophyllids, are regarded as important because of the alleged similarities in the ecology of the oncospheres of caryophyllids and pseudophyllids. Such differences indicate that the two groups are indeed different from each other and not that caryophyllids should be removed from the cestoda.

Contrasting the morphogenesis of caryophyllids and segmented cestodes, Bazitov calls attention to the fact that metamorphosis in the former is associated with degeneration of the cercomer alone and involves no other larval structures. He does not feel that this difference, that is loss of cercomer only, can be explained by the neoteny hypothesis because (Bazitov, 1976, p. 1782); ". . . one would be forced to postulate secondary changes in the type of development, occurring with typical metamorphosis, in the ancestors of caryophyllids". He believes it is more reasonable to derive them from free-living turbellarians where there is a similar type of direct development. In my opinion, neoteny or progenesis need not be concerned with ancestors because either process could have occurred late in the evolution of the cestodes.

Final evidence is in the difference between the histological organization of the parenchyma between caryophyllids and segmented cestodes, differences Bazitov considers striking. His conclusions are based on a study of the histology of *Caryophyllaeus laticeps* and *Khawia sinensis*; Will's (1893) study of the former species is not cited. Perhaps the most outstanding difference is the presence of giant basophilic and oxyphilic cells, the former are compared with the neoblasts of planarians. Other differences are cited in the sub-cuticular layer. The "faserzellen", so prominent in the neck of caryophyllids, are designated as neoblast cells; these giant cells and other "specific cellular elements" are said to be absent from segmented cestodes. A final evidence of uniqueness is the extraordinary development of the uterine glands. According to Bazitov (1976, p. 1784), "The abundance of specific features, present in the parenchyma and tegument, is in direct contradiction with current views as to the place of caryophyllids in the systems of flatworms and as to their evolution". I agree that the histological structure and especially the size of cells of caryophyllids is striking and different from that of segmented cestodes. However, should one not expect such differences between primitive and advanced forms? One might also expect them in neotenic or progenetic forms.

According to Bazitov, the neotenic hypothesis should be supported by similarities in the cleavage pattern, morphology of oncosphere, and tissue organization between caryophyllids and segmented cestodes. I agree in the first case but not in the latter two because much would depend on the temporal relationship between the parent stock and the appearance of the progenetic stage. For example, because the progenetic (one-host) and adult (non-progenetic) stage of *Archigetes* often occur in the same species, it is doubtful that there is any difference in the cleavage pattern and other features between the two forms. Where there is no one-host cycle, as is the condition with the great majority of caryophyllids, then progenesis may have been a factor in the initial evolution of a monozoic condition from a polyzoic one. Once established, this monozoic form underwent evolution and radiated into diverse genera and species. Hence one would expect great differences between the present monozoic forms and the former polyzoic ancestors. If comparisons are to be made they should be between the most comparable stages, namely the monozoic caryophyllids and plerocercoids of pseudophyllids (assumed to have given rise to the monozoic forms) and not between monozoic and polyzoic stages which are at different developmental states. It would appear to me that the differences cited by Bazitov confirm that monozoic cestodes are basically different from polyzoic ones, but they do not disprove that one could have evolved from the other.

As a final line of evidence, Bazitov does not regard the similarities of the reproductive systems (vitellaria distribution) of caryophyllids and pseudophyllids as being evidence of relatedness but as examples of convergence. The similarities in the position of the genital pores is coincidental; since there is so much variation in the genital ducts of the pseudophyllidea it is a matter of chance that some of these variations would correspond to the condition in caryophyllids. He supports his argument with Dubinina's (1974b) view that the reproductive system is of secondary importance in defining large taxonomic units; more important are the attachment organs which are not discussed by Bazitov. Although he has some suggestions regarding the systematic position of caryophyllids, he is frank to admit (p. 1785) that ". . . . this problem remains open so far, it can be solved after a comparative study of all cercomeromorph groups of flatworms". In my view, Bazitov's approach to the complexities of caryophyllid phylogeny is sound for it deals not with speculation but with aspects of basic biological relationships.

The most extensive recent review of caryophyllid phylogeny is that of Kulakovskaya and Demshin (1978); both had written shorter reviews earlier (Demshin, 1971; Kulakovskaya, 1976). This joint paper is in part a re-affirmation of each other's views, plus a response to Mamaev's (1975) critique of Demshin's earlier paper: Bazitov's (1976) long paper is not mentioned, unfortunately. It does not add any new ideas but gives a compre-hensive, but all too brief, review of past views. Some of their own ideas are expanded, for example, they believe that oligochaetes were the primary hosts of cestodes because of (a) narrow specificity of caryophyllideans in oligo-chaetes, (b) long period (4–6 months) of development in oligochaetes com-pared to short period in fish, (c) high degree of morphological differentiation in oligochaetes, and (d) wide range of infectivity, i.e. from moment of cer-comer formation to almost mature stage in oligochaetes. Bothria evolved on *Archigetes* while it was a parasite in the *intestine* of oligochaetes and were retained when caryophyllids moved to the body cavity. What prompted this habitat change is not explained, nor is there any realization that the gut of tubificids is probably too small for *Archigetes* unless it was a much smaller parasite. Furthermore, the fact that there are no mature cestodes in the gut of any invertebrate, or that the oncosphere leaves the gut shortly after hatching, would argue against an intestinal habit in oligochaetes for *Archi-getes*. Progenesis is rejected as an explanation for *Archigetes* because they admit to being unable to see the biological rationale for a parasite to move from the nutritionally rich environment of the intestine to the poorer one of the oligochaete body cavity (see Section IVC). Since they believe that oligo-chaetes were the original hosts, shortening the cycle by elimination of the most recently added vertebrate host does not appear reasonable. While the parasites of oligochaetes evolved to caryophyllideans, those of crustaceans gave rise to contemporary cestodes. The ancestors of rhabdocoela and acoela may also have given rise to the cestodes and monogeneans. They conclude by regarding the caryophyllideans as an ". . . independent group at the rank of superorder or subclass . . .". This is said to support the conclu-sions of Bazitov (1976) and Mackiewicz (1972); in the latter case the authors have clearly misinterpreted the paper.

D. COTYLODA CONCEPT

As a sequel to their influential earlier work (Wardle and McLeod, 1952), Wardle *et al.* (1974) attempt to summarize the advances in the zoology of tapeworms for the period 1950–1970. This work, like the first, contains a number of new taxa, nine new orders and one new class, and is destined to have a stimulating influence on the course of cestode classification. In their review of cestode phylogeny, they accept the hybrid scheme of Price (1967)

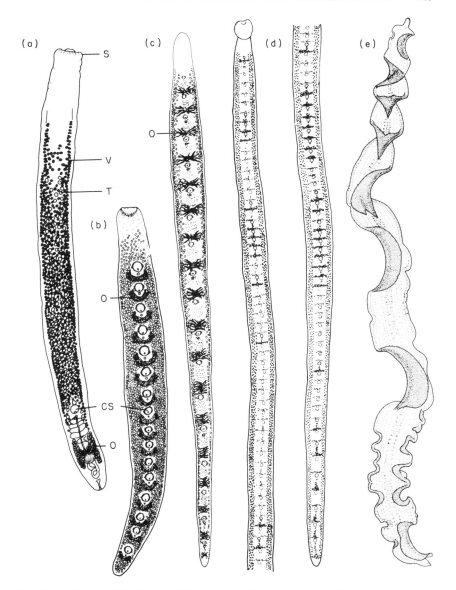

FIG. 11. (a) *Monobothrium hunteri*, (CS, cirrus sac; O, ovary; S, scolex; T, testis; vitellarium). (original). (b) *Cyathocephalus truncatus* (CS, cirrus sac; O, ovary) (original). (c) *Spathebothrium simplex* (O, ovary). (original). (d) *Bothrimonus* sp. (original). (e) *Anatrum tortum* (adapted from a photograph by Rees, 1969). Drawings are not to scale.

which is based primarily on the work of Bychowsky (1957) and Llewellyn (1965). In it, the cestodes are made an offshoot of the amphilinideans, which derive from gyrocotylideans that in turn branch from the monogeneans which evolved from the rhabdocoel turbellarians. The authors conclude that the primeval tapewor.ns were small, slender, without suckers, but with a sensory apical organ that was an invagination of the body surface that led into a loose mass of parenchyma which served as a gut. They also concluded that (p. 19) "Traces of a former alimentary tract and mouth perhaps occur in some tapeworm larval sta˷es". They are no doubt referring to the terminal invagination and frontal gland complex of some procercoids, i.e., *Diphyllobothrium*. From such an ancestor, the pseudotapeworms (Cotyloda) and true tapeworms (Eucestoda) eventually evolved.

The Cotyloda constitute a new class made up of the following six orders of pseudotapeworms: Gyrocotylidea and Amphilinidea (provisional members, Figs. 1a and 1b) and Caryophyllaeidea (Fig. 11a), Spathebothridea (Figs. 11b, 11c), Diphyllidea (Fig. 11d), and Pseudophyllidea (Fig. 11e); all the other tapeworms are relegated to the class Eucestoda. Among the characteristics of the Cotyloda are: difossate (dibothriate), segmentation weak or absent, segments (if present) produced by autotomy of an invading plerocercoid; true proglottization and apolysis lacking; life cycle with an operculated and embryonated egg, a procercoid (with cercomer) in an invertebrate, a plerocercoid (without cercomer) in a second invertebrate; and (p. 22) ". . . an adult worm in a vertebrate host, derived by instant autotomy of an invading plerocercoid representing a precociously sexual, long extinct pseudotapeworm". "We regard Cotyloda as monozootic and Eucestoda as polyzootic". Cotyloda is thus made up of essentially progenetic forms. Instead of "progenetic", the authors prefer to use "neotic" to describe difossate cestodes. Except for the Gyrocotylidea and Amphilinidea, commonly known as cestodarians and differing from the others in such basic features as a 10-hooked lycophora larva, the remaining orders were formerly subsumed as families in the order Pseudophyllidea. It is therefore not surprising that Wardle *et al.* (1974) have again united them but at the class level. There appears to be little acceptance of the Cotyloda in spite of the fact that Logacev (1970) has found that the connection between proglottids in the Pseudophyllidea was basically different from that in the Cyclophyllidea. Czaplinski (1972) earlier had used many of the ideas of Wardle and McLeod (1952) and derived the tetrafossate cestodes (Eucestoda) from Acoela and difossate ones from the Amphilinidea.

The evolution of the Caryophyllidea is similar to that of the difossate worms. According to Wardle and McLeod (1952) the difossate worms originated as endosymbionts of invertebrates from a protocestode stock that also gave rise to turbellarians. Such endosymbionts may have originated independently and many times. They lacked well-developed holdfasts and, most important, delayed autotomy (cross division of the body which produces primary segmentation) had not yet evolved. Instead, there is secondary segmentation imposed upon a plerocercoid-like form, as evident in *Ligula*. In some forms this segmentation became highly developed, e.g. *Diphyllo-*

bothrium; however, the tapeworm remains basically monozootic because proglottids are not shed. Segmentation developed to various degrees and in caryophyllids, not at all. Thus they are, in the terminology of Wardle and McLeod (1952, p. 153) "... essentially neotenic forms". More recently, Wardle *et al.* (1974, p. 28) regard caryophyllids as "... actually sexually precocious juveniles of long-extinct protocestodes". I interpret the scheme of Wardle and McLeod (1952) and Wardle *et al.* (1974) to mean that caryophyllideans and other difossate forms are progenetic because the plerocercoid-like stage never developed true segmentation and apolysis characteristic of the tetrafossate line and thus remain basically a sexually mature juvenile tapeworm. In this scheme, caryophyllideans are primarily monozoic because they never developed instant autotomy.

In my view, the Cotyloda concept (without Gyrocotylidea and Amphilinidea, however) has considerable merit because it uses fundamental developmental patterns and scolex morphology to unite forms that share so many other characteristics. With such a scheme, various degrees of progenesis and segmentation can be explained easily for various cycles (*Ligula*, *Cyathocephalus*, etc.) because all share the same developmental potential. However, it does not explain why caryophyllids are not segmented although they live in the gut, the habitat that selected for segmentation. On the other hand, is it possible that the calcareous corpuscle distribution may also reflect a form of segmentation (see Section IIIB). In my view, another important value of the Cotyloda is conceptual because it breaks the cestodes into at least two new large groups and establishes a framework for re-examining cestode classification.

E.   COMMENT

From the preceding 10 papers it is evident that there is remarkably little agreement as to the origin and evolutionary history of the caryophyllid cestodes. Differing interpretations, lack of data, lack of knowledge of earlier views, and the intrinsic complexities of dealing with hypothetical ancestral forms contribute to the confusion. Some believe caryophyllids evolved from acoels or acoelous ancestors (Demshin, 1971; Czaplinski, 1972; Malmberg, 1974) or rhabdocoels or rhabdocoel-like forms (Freeman, 1973; Wardle *et al.*, 1974; Kulakovskaya, 1976) or possibly both (Kulakovskaya and Demshin, 1978). Except for Mamaev (1975), who believes vertebrates were the first hosts, there appears to be general agreement that the first hosts were invertebrates; many believe that these invertebrates were oligochaetes (Demshin, 1971; Czaplinski, 1972; Malmberg, 1974; Jarecka, 1975; Kulakovskaya, 1976; Kulakovskaya and Demshin, 1978), and some derive all cestodes through a caryophyllid or caryophyllid-like line (Demshin, 1971; Freeman, 1973; Malmberg, 1974; Jarecka, 1975; Kulakovskaya, 1976; Kulakovskaya and Demshin, 1978). There appears to be unanimous agreement that caryophyllids are primarily monozoic even though progenesis may have played a role in their evolution (Wardle *et al.*, 1974; Mamaev, 1975). Finally, there are some who would elevate caryophyllids to the status of an independent Class or Subclass (Bazitov, 1976; Kulakovskaya and Demshin, 1978).

It is perhaps significant that so many workers now follow Llewellyn (1965) and derive the strobilate cestodes through a protocaryophyllidean or caryophyllidean line. Earlier, the origin through progenesis from a polyzoic line was widely held. Rees (1969), whose paper appears to have been overlooked by all authors cited above except Wardle et al. (1974), also develops the strobilate cestodes through a similar line. Like Llewellyn, she assumes that a protomonogenean invaded the gut of a vertebrate and evolved to the ancestral protocaryophyllidean, present today in the form of Caryophyllidea. From this monozoic ancestral stock evolved the strobilate cestodes through a series of morphological steps as follows (existing forms in parentheses): (1) genitalia repeated (Spathebothriidae, Fig. 11c) (2) elongation, better development of scolex with genital pores on one surface (Cyathocephalidae, Fig. 11b; Diplocotylidae, Fig. 11d) or uterine pore on ventral surface, cirro-vaginal pore on dorsal (*A. tortum*, Fig. 11e), (3) pseudopolysis, or shedding of exhausted posterior end as strobila gets too long to be secured in gut by unspecialized scolex (*A. tortum*), (4) scolex specialized (Pseudophyllidea), (5) dilation of uterine pore and temporary retention of eggs in uterus (Pseudophyllidea) and finally, (6) closure of uterine pore and retention of eggs in uterus, various types of apolysis (Cyclophyllidea and most other cestodes).

Because a gradually evolving series is formed, Rees (1969) concludes that it is difficult to divide it into orders and therefore (p. 536) "It might be advisable to accept Hyman's (1951) suggestion to include them all, in separate families, in the Pseudophyllidea". Such a conclusion is very much like that of Nybelin (1922), except that he also created a number of subfamilies, Caryophyllaeinae and the Cyathocephalinae. In judging the unified pseudophyllidea concept, it is important to remember that Nybelin (1922), Hyman (1951) and Rees (1969) have relied *exclusively* on morphological criteria, ignoring host-parasite relationships or life-cycle patterns. Hyman (1951, p. 421) went so far as to state that "In general, no phylogenetic importance can be attributed to the monozoic condition", after citing the series: *Archigetes–Biacetabulum–Caryophyllaeus–Ligula–Schistocephalus*. Because her basic assumption was that neotony led to the monozoic condition it is not difficult to understand why she reached that conclusion. Indeed, neotenic forms generally are not accorded any special phylogenetic position. It is precisely that point that convinces me that *Archigetes* should not be given special phylogenetic significance over that of other caryophyllids (see Section IVE) or that caryophyllideans as a group should not be regarded as stem forms although progenesis may have a role in the *early* evolution of the group. However, there is no *proof* that the monozoic condition is, as Hyman assumes, indeed a result of neoteny (her terminology). That there is a tendency for progenesis in the Pseudophyllidea is clearly evident in the reduction or absence of segmentation. But the hypothetical step from a polyzoic to a monozoic form is in no way a clear-cut example of neoteny (progenesis). If it were (there is still room for other interpretations), and such a change had a permanent genetic basis (see Section II) and was accompanied by a completely new life cycle (with oligochaetes) or group of hosts (cypriniforme fishes), and had given rise to a large (over 100 species) widely distributed group (one of the major cestode groups in freshwater fishes)—then to designate such a group of monozoic forms simply a family of the Pseudophyllidea *because* neoteny (or progenesis) *may* have occurred somewhere in the past, is to ignore the evolutionary process of mechanisms for the formation of higher taxa (see Section II 3).

VI. Discussion

Absence of a fossil record and great adaptive changes in highly specialized forms through a long evolution as a parasite have greatly complicated the analysis of cestode evolution. Any evolutionary scheme is therefore clearly speculative and in truth it is possible to arrive at different conclusions utilizing the same facts. In the final analysis, the most reasonable evolutionary scheme and systematic positioning of caryophyllids will depend on the answers to certain key questions.

1. *Are caryophyllids primarily or secondarily monozoic?*

The majority of zoological opinion favours the primary origin. However, in my view, there are a number of possibilities. For example, it is possible that the early ancestral caryophyllideans were *secondarily* derived from progenetic plerocercoids of a pseudophyllidean stock and through a change in the role of the vitelline cell nucleus (Section III C) and an alteration of the segmentation process (Section III A), were able to move from arthropods to oligochaetes as intermediate hosts. Once in oligochaetes that line gave rise to the caryophyllideans. Having separated from a polyzoic ancestor, the newly evolving caryophyllidean is primarily monozoic with vestiges of the polyzoic potential for segmentation (Section III B) and has the procercoid stage of a polyzoic ancestor. Since no strobilate stage follows, or ever followed, the plerocercoid-like stage once the oligochaete became incorporated in the cycle, it seems logical to regard caryophyllids as primarily monozoic adult cestodes and not progenetic plerocercoids. If one accepts this scenario, then caryophyllids are best regarded as a dead-end branch off a polyzoic line, perhaps the Spathebothriidea or Pseudophyllidea.

In my opinion, this interpretation has considerable merit because (a) it is the most parsimonious when viewed in the total perspective of cestode evolution, (b) it involves only those cestodes (i.e. Pseudophyllidea or relatives) that share common morphological or biological features, (c) it is consistent with known evolutionary and ecological theory regarding parasitism and requires no hypothetical ancestors, and (d) it utilizes a biological phenomenon (progenesis, see Section II) that is common in the animal kingdom. The most important parts of this scheme are the changes in vitelline cell nuclear function, from a purely regulatory role to that of glycogen storage, and possible suppression and eventual loss of the capacity to segment. In the latter case, a monozoic cestode would evolve through the combination of progenesis, with its truncation of development, and a mutation that prevents the replication of reproductive systems. A mutation could of course also explain the appearance of the nuclear vacuole which, because it conferred greater fitness on the individuals, was selected for and became fixed in the genome of the stem form. Utilization of oligochaetes was initially a chance event that was selected for (see Section IV D) and thus became associated only with the caryophyllid cycle, whereas other cestodes continued to evolve with arthropods in their cycles.

A second alternative, the reverse of the first, is that the monozoic body plan is primary and that nowhere in the evolutionary history is there any polyzoic ancestor. In this view, caryophyllideans can be (a) ancestral to polyzoic cestodes or, (b) an independent group that diverged *before* proglottization evolved.

In the case of (a), it is assumed that some caryophyllideans did not evolve while others added a strobilate stage, at the same time changing the cycle from an oligochaete to an arthropod intermediate host. From this new strobilate form, there is radiation into different cestode groups. If one accepts this alternative, then it is necessary to explain why oligochaetes are not in the life cycles of many other cestodes, why procercoids and plerocercoids are not more widely distributed in cycles, why nuclear storage nuclei are absent from all other cestodes (as far as we know) and why some caryophyllideans strobilated and others did not, yet both lived in the gut. It is difficult for me to answer these and other equally puzzling questions without postulating numerous changes that are not supported by known facts or examples. Furthermore, by placing caryophyllids at the base of cestode evolution one wonders why no other cestodes have their principal characteristic—monozoic body plan. At the very least, a stem form would be expected to give rise to other forms, rather similar to it, yet caryophyllids remain so distinct from other cestodes.

In the case of (b), separate lines of evolution for monozoic and polyzoic cestodes from a common monozoic ancestor is fraught with problems of explaining morphological similarities of eggs, reproductive systems and procercoid or plerocercoid stages between caryophyllids and pseudophyllids. So profound are these problems that using convergence as an explanation seems to me to avoid the problems rather than solving them.

2. *Are tubificids the original host of caryophyllids?*

Archigetes, of course, is often used to support the affirmative opinion. But since it is possible to cast considerable doubt on the validity of such opinions (see Section IV), the support for an invertebrate original host is weakened considerably, if not completely lost. Further erosion of the affirmative view occurs when one considers that aquatic arthropods (copepods and amphipods) and annelids (oligochaetes) generally lack sufficient size, nutrient-rich habitats, and long enough life-span to allow a flatworm to become parasitic by providing enough space, food and time to evolve a reproductive potential high enough to sustain a parasitic mode of life. The almost total absence of one-host life cycles of cestodes with arthropods or annelids would tend to support this analysis. Unlike digenetic trematodes who have molluscs as a common thread in their biology, cestodes have a great variety of intermediate hosts that include invertebrates and vertebrates, suggesting that many were added to cycles as the original hosts, vertebrates, dispersed and exposed their evolving cestodes to various other organisms. Under some circumstances, however, where there are high invertebrate populations as well as desirable morphological features (e.g. size) and long life span, tubificids became hosts (Section I C 3) for sexually mature procercoids of small caryophyllids when progenesis truncated the normal two-host cycle. In this case, the original hosts were probably freshwater fish.

VII. Conclusions
(relevant Sections in parentheses)

The relationship of the evolutionary history of caryophyllid cestodes to strobilate forms is still poorly understood. Originally placed with Cestodaria (I B), there seems little question that they are true cestodes, allied to strobilate forms through a common oncosphere stage and basic similarities of morphology. Nevertheless, caryophyllid morphology and biology is remarkably different from that of any other group of cestodes (I C). Viewed in the perspective of progenesis (II) and current ecological theory, it is possible to develop an evolutionary scheme that offers explanations for the monozoic morphology and unique cycle with oligochaetes. Recent experimental evidence (III A) that the normally polyzoic *Echinococcus* can yield monozoic forms suggests that the developmental difference between the two body forms may not be too great. Conversely, studies on the distribution of calcareous corpuscles (III B) in *Glaridacris* indicate that some monozoic forms may have traces of cryptic segmentation.

A major consequence of the monozoic morphology is that caryophyllids lack the high reproductive potential so characteristic of the cestoda as a whole. Their evolution, survival and relationship to strobilate cestodes can be tied to adaptations that often may be related to this diminished reproductive capacity. Viewed in this perspective, a knowledge of various morphological features (III) may help to explain how they have contributed to the survival and evolution of such a unique group of cestodes. From cytological studies (III D), it is clear that caryophyllids have been actively evolving and show a tendency toward polyploidy. With ploidy has come parthenogenesis (III E), itself of possible selective value because more energy can be devoted to egg production. It would appear that the presence of a large glycogen vacuole in the nucleus of vitelline cells may have had considerable influence in directing the evolution of intermediate host selection from an arthropod to oligochaete. Another possible adaptation related to energy relationships may be the high levels of vitamin B_{12} (III G) which, like parthenogenesis, may allow a greater energy allocation for egg production than for male gametes.

However, not all features appear to have apparent adaptive significance, instead they may shed light on phylogenetic relationships. The apparent absence of C-type virus particles (III H) from caryophyllids and common occurrence in pseudophyllideans is difficult to assess. More difficult to reconcile, is the observation that caryophyllid sperm, with a single axoneme, are more similar to those of cyclophyllideans than pseudophyllideans (III F) whose sperm have two axonemes.

Of all cestodes, *Archigetes* has a singular position because it can have a one-host cycle. On the basis of an extensive analysis of *Archigetes* morphology, biology and ecology (IV), it would appear that it is not an adult or ancestral cestode, is no more primitive than other caryophyllids and was probably first parasitic in a fish and secondarily in oligochaetes.

Review of various evolutionary schemes since 1971 (V) shows that there is remarkably little agreement on the origin and evolutionary history of caryophyllids. Clearly, the question of the presence of entoderm (V A) in cestodes has influenced whether or not an acoel, rather than rhabdocoel, was the ancestral stock. Strong arguments can be made for both alternatives. Perhaps most interesting is the view, premature in my opinion, that caryophyllids should be placed with the pseudophyllidea, and a few other orders, in a new class of pseudotapeworms, Cotyloda (V B); or that they should be elevated to the status of an independent Class or Subclass (V C). Although few workers go this far, others take the equally extreme view, which I cannot support, that caryophyllids should be considered a family of the Pseudophyllidea (V D). In my opinion, there is ample justification for the order status.

The elusive nature of evolutionary process prevents dogmatic conclusions. With so complex a group as the caryophyllids, it is indeed possible to suggest various evolutionary pathways. In my opinion, a strong argument (VI) can be made for the view that the early ancestral caryophyllideans were secondarily derived from a progenetic plerocercoid but evolved quickly and very early to become primarily monozoic because the strobilate stage disappeared once oligochaetes became intermediate hosts. The original hosts were probably freshwater fish.

Whatever the interpretation of the evolution and phylogenetic position of caryophyllid cestodes, no one can deny that in them we have a remarkable group of cestodes from which we can learn much of the basic principles of parasite evolution.

Acknowledgments

I would like to express my deep appreciation to a number of friends and colleagues who have aided me in various ways in the preparation of this analysis and review: Dr. B. Grabda-Kazubska, for her expert translations of many of the Russian and some of the Polish papers; Dr. Z. Kabata, for his translation of the Bazitov (1976) paper; Professors E. Scatton, P. Marfey and H. Ghiradella for assistance in the translation of portions of Russian or Ukrainian literature; Professor H. Hirsch, for assistance in the interpretation of some German literature; Professor J. D. Smyth for examples of monozoic *Echinococcus* and access to a paper in press; Dr. C. A. Hall for vitamin B_{12} assay; Professors H. Tedeschi and D. Holmes for helpful discussions on glycogen metabolism and progenesis–genome considerations, respectively; Dr. S. Mackiewicz, for editorial review of all parts of this paper; Mr. R. Loos, for making the drawings and Mr. R. Speck for photographic assistance; and finally Miss Laurie Callahan for her ever cheerful spirit in preparing the review from manuscript to final stage.

REFERENCES

Amin, O. M. (1978). On the crustacean hosts of larval acanthocephalan and cestode parasites in southwestern Lake Michigan. *Journal of Parasitology* **64**, 842–845.

Baer, J. (1952). "Ecology of Animal Parasites", University of Illinois Press, Urbana.

Bazitov, A. A. (1974). [The theory of germ layers and embryonic development of *Hymenolepis nana* (Cyclophillidea).] *Arkhiv Anatomii Gistologii i Embriologii* **66**, 80–85. (In Russian.)

Bazitov, A. A. (1976). (The status of Caryophyllidea in the system of platyhelminthes.) *Zoologicheskii Zhurnal* **55**, 1779–1787. (In Russian.)

Bazitov, A. A. and Lapkalo, A. V. (1977). [Fission and gastrulation in *Microsomacanthus paramicrosoma* (Cyclophyllidea, Cestoda).] *Parazitologiia* **11**, 104–112. (In Russian.)

Befus, A. D. and Podesta, R. B. (1976). Intestine. *In* "Ecological Aspects of Parasitology" (C. R. Kennedy, ed.), pp. 303–325, North-Holland Publishing Company, Oxford.

Beklemishev, W. N. (1969). "Principles of Comparative Anatomy of Invertebrates, Volume 2. Organology" Translated from Russian version, 3rd edition (1964) by J. M. MacLennan (Z. Kabata, ed.), Oliver and Boyd Ltd., Edinburgh.

Boyce, N. (1974). Biology of *Eubothrium salvelini* (Cestoda: Pseudophyllidea), a parasite of juvenile sockeye salmon (*Oncorhynchus nerka*) of Babine Lake, British Columbia. *Journal of Fisheries Research Board of Canada* **31**, 1735–1742.

Boyer, B. C. (1971). Regulative development in a spiralian embryo as shown by cell deletion experiments on the Acoel, *Childia*. *Journal of Experimental Zoology* **176**, 97–106.

Brinkhurst, R. O. and Jamieson, B. G. M. (1971). "Aquatic Oligochaeta of the World." University of Toronto Press, Toronto.

Burt, D. R. R. (1963). The germinal layers of cestodes. *Biological Journal* **3**, 3–7.

Burt, M. D. B. and Sandeman, I. M. (1969). Biology of *Bothrimonus* ($= Diplocotyle$) (Pseudophyllidea: Cestoda). Part I. History, description, synonymy, and systematics. *Journal of the Fisheries Research Board of Canada* **26**, 975–996.

Bychowsky, B. E. (1957). (Monogenetic trematodes, their systematics and phylogeny.) (Translation, W. J. Hargis Jr., ed.), American Institute of Biological Sciences, Washington, D.C., 1961.

Cable, R. (1971). Parthenogenesis in parasitic helminths. *American Zoologist* **11**, 267–272.

Calentine, R. (1962). *Archigetes iowensis* sp.n. (Cestoda: Caryophyllaeidae) from *Cyprinus carpio* L. and *Limnodrilus hoffmeisteri* Claparède. *Journal of Parasitology* **48**, 513–524.

Calentine, R. (1964). The life cycle of *Archigetes iowensis* (Cestoda: Caryophyllaeidae). *Journal of Parasitology* **50**, 454–458.

Calentine, R. (1965). The biology and taxonomy of *Biacetabulum* (Cestoda: Caryophyllaeidae). *Journal of Parasitology* **51**, 243–248.

Calentine, R. (1967). Larval development of four Caryophyllaeid cestodes. *Proceedings of the Iowa Academy of Science* (1965) **72**, 418–424.

Calentine, R. and DeLong, B. C. (1966). *Archigetes sieboldi* (Cestoda: Caryophyllaeidae) in North America. *Journal of Parasitology* **52**, 428–431.

Calow, P. and Jennings, J. B. (1974). Calorific values in the phylum platyhelminthes: The relationship between potential energy, mode of life and the evolution of entoparasitism. *Biological Bulletin* **147**, 81–94.

Cameron, T. W. M. (1956). "Parasites and Parasitism". John Wiley, New York.

Cameron, T. W. M. (1964). Host specificity and the evolution of helminthic parasites. *In* "Advances in Parasitology" (B. Dawes, ed.), Vol. 2, pp. 1–34. Academic Press, New York and London.

Chowdhury, N. and DeRycke, P. H. (1977). Structure, formation and functions of calcareous corpuscles in *Hymenolepis microstoma*. *Zeitschrift für Parasitenkunde* **53**, 159–169.

Coil, W. (1970). Studies on the biology of the tapeworm *Dioecocestus acotylus* with emphasis on the oogenotop. *Zeitschrift für Parasitenkunde* **33**, 314–328.

Coil, W. (1972). Studies on the dioecious tapeworm *Gyrocoelia pagollae* with emphasis on bionomics, oogenesis, and embryogenesis. *Zeitschrift für Parasitenkunde* **39**, 183–194.

Costello, D. P. and Henley, C. (1976). Spiralian development: A perspective. *American Zoologist* **16**, 277–291.

Czaplinski, B. (1972). (Problems of interrelations between the evolution of Cestoda and their hosts.) *Wiadomosci Parazytologiczne* **18**, 373–383, 421 (In Polish.)

Darlington, Jr. P. J. (1957). "Zoogeography: the geographical distribution of animals". John Wiley, New York.

DeBeer, G. (1958). "Embryos and ancestors", 3rd edition. Oxford University Press, Oxford.

Demshin, N. I. (1971). (On phylogeny of life cycles of cestodes.) *Biological and Medical Investigations in Far East*, pp. 280–296, USSR Academy of Sciences, Institute of Biology and Pedology, Far Eastern Scientific Centre, Vladivostok. (In Russian.)

Demshin, N. I. (1975). (Oligochaeta and hirudinea as intermediate hosts of helminths.) Publishing house "Nauka", Siberian Branch Novosibirsk. (In Russian.)

Demshin, N. I. (1977). (On the parasitic flatworm larval fauna in oligochaetes and leeches in the coastal region.) *Proceedings of the Institute of Biology and Pedology* (new series) **47** (150), 69–88. (In Russian.)

Demshin, N. I. (1978). [Biology of *Khawia japonensis* (Caryophyllidea, Cestoda), a parasite of the Amur carp.] *Parasitologiia* **12**, 493–496. (In Russian.)

Domenici, L. and Gremigni, V. (1974). Electron microscopical and cytochemical study of vitelline cells in the freshwater triclad *Dugesia lugubris* s.l. II. Origin and distribution of reserve materials. *Cell Tissue Research* **152**, 219–228.

Dougherty, R. M., DiStefano, H., Feller, U. and Mueller, J. F. (1975). On the nature of particles lining the excretory ducts of pseudophyllidean cestodes. *Journal of Parasitology* **61**, 1006–1015.

Dremkova, P. (1974). (Helminth larvae in oligochaetes at the Tsimlyanskii water reservoir.) *Voprosy Parazitologii Zhivotnykh Yugo-Vostoka USSR* (1974), 3–7. (In Russian.)

Dubinina, M. N. (1966). [Ligula, (Cestoda: Ligulidae) fauna USSR] Research Monograph, Publishing house, "Nauka", Moscow, Leningrad. (In Russian.)

Dubinina, M. N. (1974a). [The development of *Amphilina foliacea* (Rud.) at all stages of its life cycle and the position of Amphilinidea in the system of Platyhelminthes.] *Parazitologicheskii Sbornik* **26**, 9–38. (In Russian.)

Dubinina, M. N. (1974b). [The state and immediate goals of taxonomy of tapeworms (Cestoidea Rud., 1808).] *Parazitologiliia* **8**, 281–292. Translated from Russian, Fish and Wildlife Service, U.S. Department of the Interior, TT 76–58187.

Edwards, S. and Mueller, J. (1978). On the apparent absence of C-virus-like particles in the Caryophyllaeidae. *Journal of Parasitology* **64**, 877.

Enghoff, H. (1976). Taxonomic problems in parthenogenetic animals. *Zoologica Scripta* **5**, 103–104.

Erasmus, D. A. and Davies, T. W. (1979) *Schistosoma mansoni* and *S. haematobium:* calcium metabolism of the vitelline cell. *Experimental Parasitology* **47**, 91–106.

Euzet, L. and Mokhtar-Maamouri, F. (1976). Developpement embryonnaire de deux Phyllobothriidae (Cestoda: Tetraphyllidea). *Annales de Parasitologie Humaine et Comparée* **51**, 309–327.

Fairbairn, D. (1970). Biochemical adaptations and loss of genetic capacity in helminth parasites. *Biological Reviews* **45**, 29–72.

Featherston, D. W. (1971). *Taenia hydatigena* III. Light and electron microscope study of spermatogenesis. *Zeitschrift für Parasitenkunde* **37**, 148–168.

Franzen, A. (1956). On spermiogenesis, morphology of the spermatozoon, and biology of fertilization among invertebrates. *Zoologiska Bidrag Från Uppsala* **31**, 355–482.

Freeman, R. S. (1957). Life cycle and morphology of *Paruterina rauschi* n.sp. and *P. candelabraria* (Goeze, 1782) (Cestoda) from owls, and significance of plerocercoids in the order Cyclophyllidea. *Canadian Journal of Zoology* **35**, 349–370.

Freeman, R. S. (1973). Ontogeny of cestodes and its bearing on their phylogeny and systematics. *In* "Advances in Parasitology" (B. Dawes, ed.), Vol. 11, pp. 481–557. Academic Press, London.

Freze, V. (1965). Proteocephalata in fish, amphibians, and reptiles. *In* "Essentials of Cestodology", Vol. V. (K. I. Skrjabin, ed.). Israel Program for Scientific Translations, Jerusalem (1969).

Ginetsinskaya, T. A. (1944). (Neoteny phenomena in cestodes.) *Zoologicheskii Zhurnal* **23**, 35–42. (In Russian.)

Ginetsinskaya, T. A., Palm. U., Bessedina, V. V. and Timopheeva, T. A. (1971). (Accumulation of reserve substances in the yolk glands of platyhelminthes.) *Parazitologiia* **5**, 147–154. (In Russian.)

Gould, S. J. (1977). "Ontogeny and Phylogeny", The Belknap Press of Harvard University Press, Cambridge, Massachusetts.

Grabda-Kazubska, B. (1976). Abbreviation of the life cycles in plagiorchid trematodes. General remarks. *Acta Parasitologica Polonica* **24**, 125–141.

Greenwood, P. H., Rosen, D. E., Weitzman, S. H. and Meyers, G. S. (1966). Phyletic studies of teleostean fishes, with a provisional classification of living forms. *Bulletin of the American Museum of Natural History* **131**, 341–456.

Grey, A. J. (1979). A comparative study of the chromosomes of twenty species of caryophyllidean tapeworms. *Dissertation Abstracts International* **B40**, 590–591.

Grey, A. J. and Mackiewicz, J. S. (1974). Chromosomes of the caryophyllidean tapeworm *Glaridacris laruei*. *Experimental Parasitology* **36**, 159–166.

Grey, A. J. and Mackiewicz, J. S. (1980). Chromosomes of caryophyllidean cestodes: diploidy, triploidy and parthenogenesis in *Glaridacris catostomi*. *International Journal for Parasitology* **10**, 397–407.

Grigelis, A. (1972). [Occurrence of the procercoids of Caryophyllaeidae (Cestoda) in oligochaeta of Lithuanian SSR waters.] *Lietuvos TSR Mokslu Akademijos Darbai, C. serija*, 2(58)*t*. 1972, 71–75. (In Lithuanian.)

Halton, D. W., Stranock, S. D. and Hardcastle, A. (1974). Vitelline cell development in monogenean parasites. *Zeitschrift für Parasitenkunde* **45**, 45–61.

Hendelberg, J. (1969). On the development of different types of spermatozoa from spermatids with two flagella in the turbellaria with remarks on the ultrastructure of the flagella. *Zoologiska Bidrag Från Uppsala* **38**, 1–52.

Hendelberg, J. (1970). On the number and ultrastructure of the flagella of flatworm spermatozoa. *In* "Comparative Spermatology", Proceedings of the International Symposium (B. Baccetti, ed.), pp. 367–374. Academic Press, New York.

Hendelberg, J. (1974). Spermiogenesis, sperm morphology, and biology of fertilization in the turbellaria. *In* "Biology of the Turbellaria" (N. W. Riser and M. P. Morse, eds.), pp. 148–164. McGraw-Hill, New York.

Hendelberg, J. (1977). Comparative morphology of turbellarian spermatozoa studied by electron microscopy. *In* "The Alex Luther Centennial Symposium on Turbellaria" (T. G. Karling and M. Meinander, eds.), pp. 149–162. *Acta Zoologica Fennica* 154 Edidit Societas Pro Fauna et Flora Fennica.

Himes, M. M. and Polister, A. W. (1962). Symposium: Synthetic process in the cell nucleus. V. Glycogen accumulation in the nucleus. *Journal of Histochemistry and Cytochemistry* **10**, 175–185.

Hyman, L. H. (1951). The invertebrates: Platyhelminthes and Rhynchocoela the acoelomate Bilateria" Volume II. McGraw-Hill, New York.

Jackson, R. C. (1976). Evolution and systematic significance of polyploidy. *Annual Reviews, Ecology and Systematics* **7**, 209–234.

Janicki, C. (1930). Uber die jüngsten Zustände von *Amphilina foliacea* in der fishchleibeshole, sowie Generelles zur Auffassung des Genus *Amphilina* G. Wagen. *Zoologischer Anzeiger* **90**, 190–205.

Jarecka, L. (1960). Life cycles of tapeworms from lakes Goldapiwo and Mamry Polnocne. *Acta Parasitologica Polonica* **8**, 47–66.

Jarecka, L. (1961). Morphological adaptations of tapeworm eggs and their importance in the life cycles. *Acta Parasitologica Polonica* **9**, 409–426.

Jarecka, L. (1975). Ontogeny and evolution of cestodes. *Acta Parasitologica Polonica* **23**, 93–114.

Jennings, J. B. (1973). Symbioses in the Turbellaria and their implications in studies on the evolution of parasitism. *In* "Symbiosis in the Sea" (W. B. Vernberg, ed.), pp. 127–160. University of South Carolina Press, Columbia.

Jennings, J. B. (1974). Digestive physiology of the Turbellaria. *In* "Biology of the Turbellaria" (N. W. Riser and M. P. Morse, eds.), pp. 173–197. McGraw–Hill, New York.

Jennings, J. B. and Calow, P. (1975). The relationship between high fecundity and the evolution of entoparasitism. *Oecologia, Berl* **21**, 109–115.

Jones, A. W. (1967). "Introduction to Parasitology". Addison–Wesley Publishing Company, Reading, Massachusetts.

Jones, A. W. and Mackiewicz, J. S. (1969). Naturally occurring triploidy and parthenogenesis in *Atractolytocestus huronensis* Anthony (Cestoidea: Caryophyllidea) from *Cyprinus carpio* L. in North America. *Journal of Parasitology* **55**, 1105–1118.

Joyeux, Ch. and Baer, J. G. (1961). Class des Cestodes. *In* "Traité de Zoologie" (P. P. Grassé, ed.), pp. 347–560. Tome IV, premier fascicule, Masson et Cie, Paris.

Kennedy, C. R. (1965a). The life-history of *Archigetes limnodrili* (Yamaguti) (Cestoda: Caryophyllaeidae) and its development in the invertebrate host. *Parasitology* **55**, 427–437.

Kennedy, C. R. (1965b). Taxonomic studies on *Archigetes* Leuckart, 1878 (Cestoda: Caryophyllaeidae) *Parasitology* **55**, 439–451.

Kennedy, C. R. (1966). The life history of *Limnodrilus hoffmeisteri* Claparède (Oligochaeta, Tubificidae) and its adaptive significance. *Oikos* **17**, 158–168.

Kennedy, C. R. (1969). Tubificid oligochaetes as food of dace *Leuciscus leuciscus* (L.). *Journal of Fish Biology* **1**, 11–15.

Körting, W. (1976). Metabolism in parasitic helminths of freshwater fish. *In* "Biochemistry of Parasites and host-parasite relationships" (H. Van den Bossche, ed.), pp. 95–100. North-Holland Publishing Company, Oxford.

Kozicka, J. (1971). Cestode larvae of the family Dilepididae Fuhrmann, 1907 parasitizing fresh-water fish in Poland. *Acta Parasitologica Polonica* **19**, 81–93.

Kulakovskaya, O. P. (1962a). [Progenetic cloveheaded worms (Caryophyllaeidae, Cestoda) in the body of oligochaetes.] *Dopovidi Akademii Nauk Ukrainskoi RSR No.* 6, pp. 825–829. (In Russian).

Kulakovskaya, O. P. (1962b). [Development of Caryophyllaeidae (Cestoda) in an intermediate host.] *Zoologicheskii Zhurnal* **41**, 986–992. (In Russian.)

Kulakovskaya, O. P. (1976). [Tapeworms (Caryophyllidea: Cestoda), their origin, modern distribution and epizootic importance.] *Gosudarstvennyi Nauchino-Issledovate l'skii Institut Ozernogo i Rechnogo Rybnago Khozyaistua Izvestia, Leningrad* **105**, 76–83. (In Russian.)

Kulakovskaya, O. P. and Demshin, N. I. (1978). (Origin and phylogenetic relationships of caryophyllideans.) *In* "Problemy Gidro-Parazitologii" (A. P. Markevic, ed.), pp. 95–104. Kiev, "Naukova Dumka". (In Russian.)

Lewontin, R. C. (1965). Selection for colonizing ability. *In* "The Genetics of Colonizing Species" (H. G. Baker and G. L. Stebbins, eds.), pp. 77–91. Academic Press, New York.

Llewellyn, J. (1965). The evolution of parasitic platyhelminths. *In* "Evolution of Parasites" (A. E. R. Taylor, ed.), pp. 47–78. *Third Symposium of the British Society of Parasity.* Blackwell Scientific Publications, Oxford.

Logachev, E. D. (1968). (Morphological principles in the evolution of cestodes.) *Kemerovskii Meditsinskii Institut,* Kemerova, 8 pp. (In Russian.)

Logachev, E. D. (1970). (Phylogenetic position of cestodes, class or type?) *Akademiya Nauk Kazakhskoi SSR, Alma Ata,* 12 pp. (In Russian.)

Løvtrup, S. (1978). Book review, S. J. Gould, "Ontogeny and Phylogeny" (1977) *Systematic Zoology* **27**, 125–130.

Lumsden, R. (1965). Microtubules in the peripheral cytoplasm of cestode spermatozoa. *Journal of Parasitology* **51**, 929–931.

Lynch, J. E. (1945). Redescription of the species of *Gyrocotyle* from the ratfish, *Hydrolagus colliei* (Lay and Bennet), with notes on the morphology and taxonomy of the genus. *Journal of Parasitology* **31**, 418–446.

MacArthur, R. W. and Wilson, E. O. (1967). "The Theory of Island Biogeography". Princeton University Press, Princeton.

Mackiewicz, J. S. (1968). Vitellogenesis and eggshell formation in *Caryophyllaeus laticeps* (Pallas) and *Caryophyllaeides fennica* (Schneider) (Cestoidea: Caryophyllaeidea). *Zeitschrift für Parasitenkunde* **30**, 18–32.

Mackiewicz, J. S. (1972). Caryophyllidea (Cestoidea): A review. *Experimental Parasitology* **31**, 417–512.

Mackiewicz, J. S. (1976). Caryophyllids—the *Drosophila* of tapeworms. *The Bulletin, New Jersey Academy of Sciences* **21**, 27.

Mackiewicz, J. S. (1978). Duplication of reproductive systems in monozoic cestodes (Caryophyllidea). *Proceedings of the Helminthological Society of Washington* **45**, 28–33.

Mackiewicz, J. S. and Blair, D. (1980). *Caryoaustralus* gen.n. and *Tholophyllaeus* gen.n. (Lytocestidae) and other caryophyllid cestodes from *Tandanus* spp. (Siluriformes) in Australia. *Proceedings of the Helminthological Society of Washington* **47**, 168–178

Mackiewicz, J. S. and Ehrenpris, M. B. (1980). Calcareous corpuscle distribution in Caryophyllid cestodes: possible evidence of cryptic segmentation. *Proceedings of the Helminthological Society of Washington* **47**, 1–9.

Mackiewicz, J. S. and Jones, A. W. (1969). The chromosomes of *Hunterella nodulosa* Mackiewicz and McCrae, 1962 (Cestoidea: Caryophyllidea). *Proceedings of the Helminthological Society of Washington* **36**, 126–131.

Mackiewicz, J. S., Cosgrove, G. E. and Gude, W. D. (1972). Relationship of pathology to scolex morphology among caryophyllid cestodes. *Zeitschrift für Parasitenkunde* **39**, 233–246

Malmberg, G. (1971). On the procercoid protonephridial systems of three *Diphyllobothrium* species (Cestoda, Pseudophyllidea) and Janicki's cercomer theory. *Zoologica Scripta* **1**, 43–56

Malmberg, G. (1974). On the larval protonephridial system of *Gyrocotyle* and the evolution of cercomeromorphae (Platyhelminthes). *Zoologica Scripta* **3**, 65–81.

Mamaev, Yu. L. (1975). (Hypotheses on the origin of cestodes from "Archigetes-like ancestors", parasites of oligochaetes.) *Zoologicheskii Zhurnal* **54**, 1277–1283. (In Russian.)

Margolis, L. (1971). Polychaetes as intermediate hosts of helminth parasites of vertebrates: a review. *Journal Fisheries Research Board of Canada* **28**, 1385–1392.

Maynard Smith, J. (1978). "The Evolution of Sex". Cambridge University Press, London.

Mayr, E. (1963). "Animal Species and Evolution". Belknap Press of Harvard University Press, Cambridge, Massachusetts.

Mead, R. and Olsen, O. (1971). The life cycle and development of *Ophiotaenia filaroides* (La Rue, 1909) (Proteocephala: Proteocephalidae.) *Journal of Parasitology* **57**, 869–874.

Mokhtar-Maamouri, F. and Swiderski, Z. (1975). Étude en microscopic électronique de la spermatogenèse de deux cestodes *Acanthobothrium filicolle benedenii* Loennberg, 1889 et *Onchobothrium uncinatum* (Rud., 1819) (Tetraphyllidea, Onchobothriidae). *Zeitschrift für Parasitenkunde* **47**, 269–281.

Mokhtar-Maamouri, F. and Swiderski, Z. (1976a). Vitellogenese chez *Echeneibothrium beauchampi* Euzet, 1959 (Cestoda: Tetraphyllidea, Phyllobothriidae). *Zeitschrift für Parasitenkunde* **50**, 293–302.

Mokhtar-Maamouri, F. and Swiderski, Z. (1976b). Ultrastructure du spermatozoide d'un cestode Tetraphyllidea Phyllobothriidae: *Echeneibothrium beauchampi*, Euzet, 1959. *Annales de Parasitologie Humaine et Comparée* **51**, 673–674.

Morseth, D. J. (1969). Sperm tail fine structure of *Echinococcus granulosus* and *Dicrocoelium dendriticum*. *Experimental Parasitology* **24**, 47–53.

Motomura, I. (1928). [Development of *Archigetes appendiculatus* (Katzel).] (Abstract) *Dobutsgaku Zasshi* **40**, 844. (In Japanese.)

Motomura, I. (1929). On the early development of monozoic cestode, *Archigetes appendiculatus*, including the oogenesis and fertilization. *Annotationes Zoologicae Japanenses* **12**, 109–129.

Mueller, J. F. (1953). Some observations on the problem of symmetry and individuality in *Taenia pisiformis*. *In* "Thapar Commemoration Volume" (J. Dayal and K. S. Singh, eds.), pp. 217–222. University of Lucknow, Lucknow, India.

Mueller, J. F. and Strano, A. J. (1974). *Sparaganum proliferum*, a sparganum infected with a virus? *Journal of Parasitology* **60**, 15–19.

Noble, E. R. and Noble, G. A. (1971). "Parasitology the Biology of Animal Parasites", 3rd edition. Lea and Febiger, Philadelphia.

Nybelin, O. (1922). Anatomish-systematische Studien über Pseudophyllideen. *Göteborgs Kungl. Vetenskaps-och. Viterhets-Samhalles Handlingar, Fjärde Följden* **26**, 228 pp.

Nybelin, O. (1962). Zur *Archigetes*-frage. *Zoologisk Bidrag, Uppsala* **35**, 292–306.

Ogren, R. E. (1956). Development and morphology of the oncosphere of *Mesocestoides corti*, a tapeworm of mammals. *Journal of Parasitology* **42**, 414–428.

Palmer, M. F. (1968). Aspects of the respiratory physiology of *Tubifex tubifex* in relation to its ecology. *Journal of Zoology, London* **154**, 463–473.

Price, C. E. (1967). The phylum platyhelminthes: A revised classification. *Rivista di Parassitologia* **28**, 249–260.

Price, P. W. (1977). General concepts on the evolutionary biology of parasites. *Evolution* **31**, 405–420.

Race, G. J., Martin, J. H., Larsh, Jr., J. E. and Esch, G. W. (1966). A study of the adult stage of *Taenia multiceps* (*Multiceps serialis*) by electron microscopy. *Journal Elisha Mitchell Scientific Society* **82**, 44–57.

Rausch, R., Scott, E. and Rausch, V. (1967). Helminths in eskimos in western Alaska, with particular reference to *Diphyllobothrium* infection and anaemia. *Transactions of the Royal Society of Tropical Medicine and Hygiene* **61**, 351–357.

Rees, G. (1969). Cestodes from Bermuda fishes and an account of *Acompsocephalum tortum* (Linton, 1905) gen. nov. from the lizard fish *Synodus intermedius* (Agassiz). *Parasitology* **59**, 519–548.

Rosario, B. (1964). An electron microscope study of spermatogenesis in cestodes. *Journal of Ultrastructure Research* **11**, 412–427.

Rosen, F. (1918). Recherches sur le développement des cestodes. I. Le cycle évolutif des Bothriocéphales. Étude sur l'origine des cestodes et leurs états larvaires. *Bulletin de la Société Neuchâteloise des Sciences Naturelles* **43**, 1–64.

Rybicka, K. (1966). Embryogenesis in cestodes. *In* "Advances in Parasitology" (Ben Dawes, ed.), Vol. 4, pp. 107–186. Academic Press, New York and London.

Schauinsland, H. (1885). Die embryonale Entwicklung der Bothriocephalen. *Jenaische Zeitschrift für Medizin und Naturwissenschaft* **19**, 520–572.

Schmidt, G. D. and Roberts, L. S. (1977). "Foundations of Parasitology", C. V. Mosby Company, Saint Louis.

Sekutowicz, St. (1934). Untersuchungen zur Entwicklung und Biologie von *Caryophyllaeus laticeps* (Pall.). *Mémoires de l'Académie Polonaise des Sciences et des Lettres, Série B, Sciences Naturelles* **6**, 11–26.

Severtsov, A. N. (1945). (The contemporary problems of evolutionary theory. Degeneration of organs and substitution.) The Collected Works. Vol. 3M. 1945. (Cited by Logachev, 1970; in Russian.)

Shinde, G. and Chincholikar, L. (1977). *Mastocembellophyllaeus nandedensis* (Cestoda: Cestodaria Monticelli, 1892) n.g. et n.sp. from a freshwater fish at Nanded, M.S., India. *Rivista di Parassitologia* **38**, 171–175.

Simkiss, K. (1976). Intracellular and extracellular routes in biomineralization. *In* "Calcium in Biological Systems" (C. J. Duncan, ed.), pp. 423–446. Symposium Society Experimental Biology XXX, Cambridge University Press, London.

Smyth, J. D. (1956). Studies on tapeworm physiology. IX. A histochemical study of egg-shell formation in *Schistocephalus solidus* (Pseudophyllidea). *Experimental Parasitology* **5**, 519–540.

Smyth, J. D. (1975). The monozoic phenomena in *Echinococcus*. *Proceedings of Second European Multicolloquy of Parasitology, Trogir*, pp. 285–291.

Smyth, J. D. (1976). "Introduction to animal parasitology." 2nd edition, Hodder and Stoughton, London.

Smyth, J. D. (1979). An *in vitro* approach to taxonomic problems in trematodes and cestodes, especially *Echinococcus*. In "Problems in the identification of parasites and their vectors" (A. E. R. Taylor and R. Muller, eds.), pp. 75–101. *Symposia of the British Society of Parasitology XVIII*, Blackwell Scientific Publications, Oxford.

Smyth, J. D. and Davies, Z. (1975). *In vitro* suppression of segmentation in *Echinococcus multilocularis* with morphological transformation of protoscoleces into monozoic adults. *Parasitology* **71**, 125–135.

Southwell, T. and Prashad, B. (1918). Methods of asexual and parthenogenetic reproduction in cestodes. *Journal of Parasitology* **4**, 122–129.

Southwell, T. and Prashad, B. (1923). A further note on *Ilisha parthenogenetica* Southwell and Prashad, 1918, a cestode parasite of the Indian shad. *Records of the Indian Museum* **25**, 197–198.

Spengel, J. W. (1905). Die Monozootie der Cestoden. *Zeitschrift für Wissenshaftliche Zoologie* **28**, 252–287.

Stephenson, J. (1930). "The Oligochaeta". Clarendon Press, Oxford.

Stunkard, H. W. (1975). Life-histories and systematics of parasitic flatworms. *Systematic Zoology* **24**, 378–385.

Sun, C. N. (1972). The fine structure of sperm tail of cotton rat tapeworm, *Hymenolepis diminuta*. *Cytobiologie* **6**, 382–386.

Suomalainen, E. (1962). Significance of parthenogenesis in the evolution of insects. *Annual Review of Entomology* **7**, 349–366.

Suomalainen, E., Saura, A. and Lokki, J. (1976). Evolution of parthenogenetic insects. In "Evolutionary Biology" (M. K. Hecht, W. C. Steere and B. Wallace, eds.), pp. 209–259, Vol. 9. Plenum Press, New York.

Swiderski, Z. (1968). The fine structure of the spermatozoon of sheep tapeworm, *Monieza expansa* (Rud., 1810) (Cyclophyllidea, Anoplocephalidae). *Zoologica Poloniae* **18**, 475–486.

Swiderski, Z. (1970). An electron microscope study of spermatogenesis in cyclophyllidean cestodes with emphasis on the comparison of fine structure of mature spermatozoa. *Journal of Parasitology* **56**, 337.

Swiderski, Z. (1976). Fine structure of the spermatozoan of *Lacistorhynchus tenuis* (Cestoda, Trypanorhyncha). *Sixth European Congress on Electron Microscopy, Jerusalem* 1976, pp. 309–310.

Swiderski, Z. and Eklu-Natey, R. (1978). Fine structure of the spermatozoon of *Proteocephalus longicollis* (Cestoda, Proteocephalidea). *Proceedings of the Ninth International Congress on Electron Microscopy, Toronto, II*, pp. 572–573.

Swiderski, Z. and Mackiewicz, J. S. (1976a). Electron microscope study of vitellogenesis in *Glaridacris catostomi* (Cestoidea: Caryophyllidea). *International Journal for Parasitology* **6**, 61–73.

Swiderski, Z. and Mackiewicz, J. S. (1976b). Fine structure of the spermatozoon of *Glaridacris catostomi* (Cestoidea, Caryophyllidea). *Proceedings of the Sixth European Congress on Electron Microscopy, Jerusalem*, 307–308.

Swiderski, Z. and Mokhtar, F. (1974). Étude de la vitellogenèse de *Bothriocephalus clavibothrium* Ariola, 1899 (Cestoda: Pseudophyllidea). *Zeitschrift für Parasitenkunde* **43**, 135–149.

Swiderski, Z., Huggel, H. and Schönenberger, N. (1970). Comparative fine structure of vitelline cells in cyclophyllidean cestodes. *Septième Congrès International de Microscopie Électronique*, Grenoble, 825–826.

Swiderski, Z., Eklu-Natey, R., Subilia, L. and Huggel, H. (1978). Fine structure of the vitelline cells in the cestode *Proteocephalus longicollis* (Proteocephalidea). *Proceedings of the Ninth International Congress on Electron Microscopy, Toronto*, **II**, 442–443.

Tkachuck, R. D., Weinstein, P. and Mueller, J. (1976). Comparison of the uptake of vitamin B_{12} by *Spirometra mansonoides* and *Hymenolepis diminuta* and the functional groups of B_{12} analogs affecting uptake. *Journal of Parasitology* **62**, 94–101.

Tkachuck, R., Saz, H., Weinstein, P., Finnegan, K. and Mueller, J. (1977). The presence and possible function of methylmalonyl CoA mutase and propoinyl CoA carboxylase in *Spirometra mansonoides*. *Journal of Parasitology* **63**, 769–774.

Tomlinson, J. (1966). The advantages of hermaphroditism and parthenogenesis. *Journal of Theoretical Biology* **11**, 54–58.

Tötterman, G. (1976). On the pathogenesis of pernicious tapeworm anaemia. *Annals of Clinical Research* **8**, Suppl. **18**, 1–48.

Tyler, S. and Rieger, R. M. (1975). Uniflagellate spermatozoa in Nemertoderma (Turbellaria) and their phylogenetic significance. *Science* **188**, 730–732.

Vergeer, T. (1936). The eggs and coracidia of *Diphyllobothrium latum*. *Papers of the Michigan Academy of Science* **21**, 715–726.

Vijayaraghavan, S. and Subramanyam, S. (1977). Chromosome number of the cestode *Lytocestus indicus*. *Current Science* **46**, 312–313.

Vik, R. (1957). Studies of the helminth fauna of Norway. I. Taxonomy and ecology of *Diphyllobothrium norvegicum* n.sp. and the plerocercoid of *Diphyllobothrium latum* (L.). *Meddelelser fra det Zoologiske Museum, Oslo Nr.* 67, pp. 25–107.

Von Bonsdorff, C.-H. and Telkka, A. (1965). The spermatozoon flagella in *Diphyllobothrium latum* (fish tapeworm). *Zeitschrift für Zellforschung* **66**, 643–648.

von Brand, T. (1973). "Biochemistry of Parasites", 2nd edition. Academic Press, New York.

Wardle, R. A. and McLeod, J. A. (1952). "The Zoology of Tapeworms". The University of Minnesota Press, Minneapolis.

Wardle, R. A., McLeod, J. A. and Radinovsky, S. (1974). "Advances in the Zoology of Tapeworms, 1950–1970". University of Minnesota Press, Minneapolis.

Weinstein, P. P. and Mueller, J. F. (1970). Contrast in vitamin B_{12} content between pseudophyllidean and cyclophyllidean tapeworms. *Journal of Parasitology* **56**, 363.

White, M. J. D. (1973). "Animal Cytology and Evolution", 3rd edition. Cambridge University Press, Cambridge.

White, M. J. D. (1978). "Modes of Speciation". W. H. Freeman and Company, San Francisco.

Will, H. (1893). Anatomie von *Caryophyllaeus mutabilis*. Rud. Ein Beitrag zur Kenntnis der Cestoden. *Zeitschrift für Wissenschaftliche Zoologie* **56**, 1–39.

Willers, W. B. and Olsen, O. W. (1969). The morphology of *Gastrotaenia cygni* Wolffhügel, 1938 (Cestoda: Aporidea) with a redescription of the genus. *Journal of Parasitology* **55**, 1004–1011.

Williams, D. D. (1978). Larval development of *Glaridacris vogei* (Cestoda: Caryophyllaeidae). *Proceedings of the Helminthological Society of Washington* **45**, 142–143.

Williams, D. D. (1979). Seasonal incidence of *Glaridacris laruei* and *G. catostomi* in Red Cedar River, Wisconsin *Catostomus commersoni. Iowa State Journal of Research* **53**, 311–316.

Williams, D. D. (1980). Procercoid development of *Isoglaridacris wisconsinensis* (Cestoda: Caryophyllaeidae). *Proceedings of the Helminthological Society of Washington* **47**, 138–139.

Wisniewski, L. W. (1930). Das Genus *Archigetes* R. Leuck. Eine Studie zur Anatomie, Histogenese, Systematik und Biologie. *Memories de L'Académie Polonaise des Sciences et des Lettres, Classe des Sciences Mathématiques et Naturelles, Série B, Sciences Naturelles* **2**, 1–160.

Wisniewski, L. W. (1932). *Cyathocephalus truncatus* Pallas II. Allgemeine Morphologie. *Bulletin International de L'Académie des Sciences de Cracovie, Classes des Sciences Mathématiques et Naturelles, Série B, Sciences Naturelles II* **2**, 311–327.

Yamaguti, S. (1959). "Systema Helminthum, The Cestodes of Vertebrates", Vol. II. Interscience Publisher, New York.

Zimmermann, H.-P., Granzow, V. and Granzow, C. (1976). Nuclear glycogen synthesis in Ehrlich ascites cells. *Journal of Ultrastructure Research* **54**, 115–123.

Subject Index

E. versicolor, interspecific relationships, 50

Esox lucius, gill parasites, 60

Eubrachiella gaini, morphology, 6

Eubothrium salvelini, calcareous corpuscles, 153

Eucestoda, classification, 140, 190

Eudactylina, attachment mechanism, 41

E. similis, nauplius stages, 27

Eudactylinidae
  attachment appendages, 38
  classification, 14
  life cycle, 27
  morphology, 6

Eudactylinodes nigra, nauplius stages, 27

Euryphorus, morphology, 6

E. brachypterus, interspecific relationships, 49, 50

E. nordmanni
  host selection, 32
  interspecific relationships, 49–50

Evolution, caryophyllid cestodes, 148
  Archigetes, 169–175

Evolutionary schemes, cestodes, 175–193, 196

Evolutionary significance, caryophyllid cestodes, 139–168

Experimental monozoic forms, caryophyllid cestodes, 151–153

F

Fasciola hepatica, vitelline cells, 157

Features, caryophyllid cestodes, 139–168

Fecampia, habits, 174

Fecundity
  caryophyllid cestodes, 172
  copepod parasites, 53–54

Feeding, parasitic copepods, 55–57

Feeding habits, caryophyllid cestode definitive hosts, 148

Feeding mode, cestodes, 176

Female reproductive tract, acanthocephalans, 84–87, 92–99
  efferent duct system, 97–99
  ligament sacs, 95
  ovarian tissue, 92, 94–97
  uterine bell, **93**

Fertilization, acanthocephalans, 119–122, 129

Fessisentis fessus
  bursa, 91
  copulatory bursa, **116**
  efferent duct system, **93**, 98, 99
  female reproductive tract, **93**, 117
  male reproductive tract, **89**
  posterior end, 77, **94**
  shelled acanthors, 124
  uterine bell, 99

F. necturorum
  ovarian balls, 86
  sex ratio, 80

F. vancleavei
  male reproductive tract, **88**
  shelled acanthors, 124
  testes, 87

Filicollis anatis, proboscis, 77

Fish
  acanthocephalan parasites, 101
  caryophyllid cestodes, 142, 146,. 147, 159,
  copepod parasites, 28, 50

Fossil copepods, fish hosts, 17, 19

Fossil fish, copepod parasites, 17, 19

Freshwater fish
  caryophyllid cestode parasites, 142, 146–148
    zoogeographical distribution, 147
  copepod parasites, 28

Frontal filaments, parasitic copepods, 39

Fundulus heteroclitus, copepod parasites, 61

G

Gadus morrhua
  copepod parasites, 32, 49
    debilitating effects of, 42, 45, 46
    salinity tolerance, 62
  oxygen consumption, 45–46

Gametogenesis, acanthocephalans, 100–113
  spermatozoa, 100–108
  oocytes, 108–113

Gammarus, caryophyllid cestode parasites, 143

Gasterosteus aculeatus, copepod parasites, 31, 32, 36

3 5282 00232 9061